A Student's Guide to Lagrangians and Hamiltonians

A concise but rigorous treatment of variational techniques, focusing primarily on Lagrangian and Hamiltonian systems, this book is ideal for physics, engineering and mathematics students.

The book begins by applying Lagrange's equations to a number of mechanical systems. It introduces the concepts of generalized coordinates and generalized momentum. Following this, the book turns to the calculus of variations to derive the Euler–Lagrange equations. It introduces Hamilton's principle and uses this throughout the book to derive further results. The Hamiltonian, Hamilton's equations, canonical transformations, Poisson brackets and Hamilton–Jacobi theory are considered next. The book concludes by discussing continuous Lagrangians and Hamiltonians and how they are related to field theory.

Written in clear, simple language, and featuring numerous worked examples and exercises to help students master the material, this book is a valuable supplement to courses in mechanics.

PATRICK HAMILL is Professor Emeritus of Physics at San Jose State University. He has taught physics for over 30 years, and his research interests are in celestial mechanics and atmospheric physics.

A Student's Guide to Lagrangians and Hamiltonians

PATRICK HAMILL
San Jose State University

CAMBRIDGE
UNIVERSITY PRESS

CAMBRIDGE
UNIVERSITY PRESS

University Printing House, Cambridge CB2 8BS, United Kingdom

One Liberty Plaza, 20th Floor, New York, NY 10006, USA

477 Williamstown Road, Port Melbourne, VIC 3207, Australia

314-321, 3rd Floor, Plot 3, Splendor Forum, Jasola District Centre, New Delhi - 110025, India

79 Anson Road, #06-04/06, Singapore 079906

Cambridge University Press is part of the University of Cambridge.

It furthers the University's mission by disseminating knowledge in the pursuit of education, learning and research at the highest international levels of excellence.

www.cambridge.org
Information on this title: www.cambridge.org/9781107617520

First published 2014
4th printing 2018

A catalogue record for this publication is available from the British Library

Library of Congress Cataloging in Publication data
Hamill, Patrick.
A student's guide to Lagrangians and Hamiltonians / Patrick Hamill.
pages   cm
Includes bibliographical references.
ISBN 978-1-107-04288-9 (Hardback) – ISBN 978-1-107-61752-0 (Paperback)
1. Mechanics, Analytic–Textbooks.   2. Lagrangian functions–Textbooks.
3. Hamiltonian systems–Textbooks.   I. Title.
QA805.H24 2013
515'.39–dc23   2013027058

ISBN 978-1-107-04288-9 Hardback
ISBN 978-1-107-61752-0 Paperback

Contents

v

Introduction

The purpose of this book is to give the student of physics a basic overview of Lagrangians and Hamiltonians. We will focus on what are called variational techniques in mechanics. The material discussed here includes only topics directly related to the Lagrangian and Hamiltonian techniques. It is not a traditional graduate mechanics text and does not include many topics covered in texts such as those by Goldstein, Fetter and Walecka, or Landau and Lifshitz. To help you to understand the material, I have included a large number of easy exercises and a smaller number of difficult problems. Some of the exercises border on the trivial, and are included only to help you to focus on an equation or a concept. If you work through the exercises, you will better prepared to solve the difficult problems. I have also included a number of worked examples. You may find it helpful to go through them carefully, step by step.

Acknowledgements

I would like to acknowledge the students in my graduate mechanics classes whose interest in analytical mechanics were the inspiration for writing this book. I also wish to acknowledge my colleagues in the Department of Physics and Astronomy at San Jose State University, especially Dr. Alejandro Garcia and Dr. Michael Kaufman, from whom I have learned so much. Finally, I acknowledge the helpful and knowledgeable editors and staff at Cambridge University Press for their support and encouragement.

PART I

Lagrangian mechanics

PART I

Lagrangian mechanics

1

Fundamental concepts

This book is about Lagrangians and Hamiltonians. To state it more formally, this book is about the variational approach to analytical mechanics. You may not have been exposed to the calculus of variations, or may have forgotten what you once knew about it, so I am not assuming that you know what I mean by, "the variational approach to analytical mechanics." But I think that by the time you have worked through the first two chapters, you will have a good grasp of the concept.

We begin with a review of introductory concepts and an overview of background material. Some of the concepts presented in this chapter will be familiar from your introductory and intermediate mechanics courses. However, you will also encounter several new concepts that will be useful in developing an understanding of advanced analytical mechanics.

1.1 Kinematics

A particle is a material body having mass but no spatial extent. Geometrically, it is a point. The position of a particle is usually specified by the vector **r** from the origin of a coordinate system to the particle. We can assume the coordinate system is *inertial* and for the sake of familiarity you may suppose the coordinate system is Cartesian. See Figure 1.1.

The velocity of a particle is defined as the time rate of change of its position and the acceleration of a particle is defined as the time rate of change of its velocity. That is,

$$\mathbf{v} = \frac{d\mathbf{r}}{dt} = \dot{\mathbf{r}}, \tag{1.1}$$

3

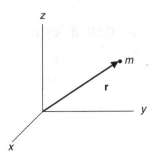

Figure 1.1   The position of a particle is specified by the vector **r**.

and

$$\mathbf{a} = \frac{d\mathbf{v}}{dt} = \ddot{\mathbf{r}}. \tag{1.2}$$

The equation for **a** as a function of **r**, **ṙ**, and t is called the equation of motion. The equation of motion is a second-order differential equation whose solution gives the position as a function of time, $\mathbf{r} = \mathbf{r}(t)$. The equation of motion can be solved numerically for any reasonable expression for the acceleration and can be solved analytically for a few expressions for the acceleration. You may be surprised to hear that the only general technique for solving the equation of motion is the procedure embodied in the Hamilton–Jacobi equation that we will consider in Chapter 6. All of the solutions you have been exposed to previously are special cases involving very simple accelerations.

Typical problems, familiar to you from your introductory physics course, involve falling bodies or the motion of a projectile. You will recall that the projectile problem is two-dimensional because the motion takes place in a plane. It is usually described in Cartesian coordinates.

Another important example of two-dimensional motion is that of a particle moving in a circular or elliptic path, such as a planet orbiting the Sun. Since the motion is planar, the position of the body can be specified by two coordinates. These are frequently the plane polar coordinates (r, θ) whose relation to Cartesian coordinates is given by the *transformation equations*

$$x = r \cos \theta,$$
$$y = r \sin \theta.$$

This is an example of a *point transformation* in which a point in the xy plane is mapped to a point in the $r\theta$ plane.

You will recall that in polar coordinates the acceleration vector can be resolved into a *radial* component and an *azimuthal* component

$$\mathbf{a} = \ddot{\mathbf{r}} = a_r\hat{\mathbf{r}} + a_\theta\hat{\boldsymbol{\theta}}. \tag{1.3}$$

Exercise 1.1 A particle is given an impulse which imparts to it a velocity v_0. It then undergoes an acceleration given by $a = -bv$, where b is a constant and v is the velocity. Obtain expressions for $v = v(t)$ and $x = x(t)$. (This is a one-dimensional problem.) Answer: $x(t) = x_0 + (v_0/b)\left(1 - e^{-bt}\right)$.

Exercise 1.2 Assume the acceleration of a body of mass m is given by $a = -(k/m)x$. (a) Write the equation of motion. (b) Solve the equation of motion. (c) Determine the values of the arbitrary constants (or constants of integration) if the object is released from rest at $x = A$. (The motion is called "simple harmonic.") Answer (c): $x = A\sin\left(\sqrt{k/m}\,t + \pi/2\right) = A\cos\sqrt{k/m}\,t$.

Exercise 1.3 In plane polar coordinates, the position is given by $\mathbf{r} = r\hat{\mathbf{r}}$. Obtain expressions for the velocity and acceleration in terms of $r, \theta, \hat{\mathbf{r}}, \hat{\boldsymbol{\theta}}$. (Hint: Express $\hat{\mathbf{r}}$ and $\hat{\boldsymbol{\theta}}$ in terms of $\hat{i}$ and $\hat{j}$.) Answer: $\mathbf{a} = (\ddot{r} - r\dot{\theta}^2)\hat{\mathbf{r}} + (r\ddot{\theta} + 2\dot{r}\dot{\theta})\hat{\boldsymbol{\theta}}$.

1.2 Generalized coordinates

In the preceding section we stated that the position of the particle was given by the vector $\mathbf{r}$. We assumed an inertial Cartesian coordinate system in which the components of $\mathbf{r}$ were (x, y, z). Of course, there are many other ways we could have specified the position of the particle. Some ways that immediately come to mind are to give the components of the vector $\mathbf{r}$ in cylindrical (ρ, ϕ, z) or in spherical coordinates (r, θ, ϕ). Obviously, a particular problem can be formulated in terms of many different sets of coordinates.

In three-dimensional space we need three coordinates to specify the position of a single particle. For a system consisting of two particles, we need six coordinates. A system of N particles requires $3N$ coordinates. In Cartesian coordinates, the positions of two particles might be described by the set of numbers $(x_1, y_1, z_1, x_2, y_2, z_2)$. For N particles the positions of all the particles are

$$(x_1, y_1, z_1, x_2, y_2, z_2, \ldots, x_N, y_N, z_N).$$

As you know, some problems are more easily solved in one coordinate system and some are more easily solved in another. To avoid being specific about the coordinate system we are using, we shall denote the coordinates by q_i and the corresponding velocities by $\dot{q}_i$. We call q_i the "generalized coordinates." Of course, for any particular problem you will chose an appropriate set of coordinates; in one problem you might use spherical coordinates in which case as i ranges from 1 to 3, the coordinates q_i take on the values r, θ, ϕ, whereas in another problem you might use cylindrical coordinates and $q_i = \rho, \phi, z$. To convert from Cartesian coordinates to generalized coordinates, you need to know the transformation equations, that is, you need to know the relations

$$q_1 = q_1(x_1, y_1, z_1, x_2, y_2, \ldots, z_N, t) \tag{1.4}$$
$$q_2 = q_2(x_1, y_1, z_1, x_2, y_2, \ldots, z_N, t)$$
$$\vdots$$
$$q_{3N} = q_{3N}(x_1, y_1, z_1, x_2, y_2, \ldots, z_N, t).$$

The inverse relations are also called transformation equations. For a system of N particles we have

$$x_1 = x_1(q_1, q_2, \ldots, q_{3N}, t) \tag{1.5}$$
$$y_1 = y_2(q_1, q_2, \ldots, q_{3N}, t)$$
$$\vdots$$
$$z_N = z_N(q_1, q_2, \ldots, q_{3N}, t).$$

It is often convenient to denote all the Cartesian coordinates by the letter x. Thus, for a single particle, (x, y, z) is written (x_1, x_2, x_3) and, for N particles, $(x_1, y_1, \ldots, z_N)$ is expressed as $(x_1, \ldots, x_n)$, where $n = 3N$.

We usually assume that given the transformation equations from the qs to the xs, we can carry out the inverse transformation from the xs to the qs, but this is not always possible. The inverse transformation is possible if the Jacobian determinant of Equations (1.4) is not zero. That is,

$$\frac{\partial(q_1, q_2, \ldots, q_n)}{\partial(x_1, x_2, \ldots, x_n)} = \begin{vmatrix} \partial q_1/\partial x_1 & \partial q_1/\partial x_2 & \cdots & \partial q_1/\partial x_n \\ & & \vdots & \\ \partial q_n/\partial x_1 & \partial q_n/\partial x_2 & \cdots & \partial q_n/\partial x_n \end{vmatrix} \neq 0.$$

The generalized coordinates are usually assumed to be *linearly independent*, and thus form a minimal set of coordinates to describe a problem. For example, the position of a particle on the surface of a sphere of radius a can be described in terms of two angles (such as the longitude and latitude). These two angles form a minimal set of linearly independent coordinates. However, we could also describe the position of the particle in terms of the three Cartesian coordinates, x, y, z. Clearly, this is not a minimal set. The reason is that the Cartesian coordinates are not all independent, being related by $x^2 + y^2 + z^2 = a^2$. Note that given x and y, the coordinate z is determined. Such a relationship is called a "constraint." We find that each equation of constraint reduces by one the number of independent coordinates.

Although the Cartesian coordinates have the property of being components of a vector, this is not necessarily true of the generalized coordinates. Thus, in our example of a particle on a sphere, two angles were an appropriate set of generalized coordinates, but they are not components of a vector. In fact, generalized coordinates need not even be *coordinates* in the usual sense of the word. We shall see later that in some cases the generalized coordinates can be components of the momentum or even quantities that have no physical interpretation.

When you are actually solving a problem, you will use Cartesian coordinates or cylindrical coordinates, or whatever coordinate system is most convenient for the particular problem. But for theoretical work one nearly always expresses the problem in terms of the generalized coordinates $(q_1, \ldots, q_n)$. As you will see, the concept of generalized coordinates is much more than a notational device.

Exercise 1.4 Obtain the transformation equations for Cartesian coordinates to spherical coordinates. Evaluate the Jacobian determinant for this transformation. Show how the volume elements in the two coordinate systems are related to the Jacobian determinant. Answer: $d\tau = r^2 \sin\theta\, dr\, d\theta\, d\phi$.

1.3 Generalized velocity

As mentioned, the position of a particle at a particular point in space can be specified either in terms of Cartesian coordinates (x_i) or in terms of generalized coordinates (q_i). These are related to one another through a transformation equation:

$$x_i = x_i(q_1, q_2, q_3, t).$$

Note that x_i, which is one of the three Cartesian coordinates of a specific particle, depends (in general) on *all* the generalized coordinates.

We now determine the components of velocity in terms of generalized coordinates.

The velocity of a particle in the x_i direction is v_i and by definition

$$v_i \equiv \frac{dx_i}{dt}.$$

But x_i is a function of the qs, so using the chain rule of differentiation we have

$$v_i = \sum_{k=1}^{3} \frac{\partial x_i}{\partial q_k} \frac{dq_k}{dt} + \frac{\partial x_i}{\partial t} = \sum_k \frac{\partial x_i}{\partial q_k} \dot{q}_k + \frac{\partial x_i}{\partial t}. \qquad (1.6)$$

We now obtain a very useful relation involving the partial derivative of v_i with respect to $\dot{q}_j$. Since a mixed second-order partial derivative does not depend on the order in which the derivatives are taken, we can write

$$\frac{\partial}{\partial \dot{q}_j} \frac{\partial x_i}{\partial t} = \frac{\partial}{\partial t} \frac{\partial x_i}{\partial \dot{q}_j}.$$

But $\frac{\partial x_i}{\partial \dot{q}_j} = 0$ because x does not depend on the generalized velocity $\dot{q}$. Let us take the partial derivative of v_i with respect to $\dot{q}_j$ using Equation (1.6). This yields

$$\frac{\partial v_i}{\partial \dot{q}_j} = \frac{\partial}{\partial \dot{q}_j} \sum_k \frac{\partial x_i}{\partial q_k} \dot{q}_k + \frac{\partial}{\partial \dot{q}_j} \frac{\partial x_i}{\partial t}$$

$$= \frac{\partial}{\partial \dot{q}_j} \sum_k \frac{\partial x_i}{\partial q_k} \dot{q}_k + 0$$

$$= \sum_k \left(\frac{\partial}{\partial \dot{q}_j} \frac{\partial x_i}{\partial q_k} \right) \dot{q}_k + \sum_k \frac{\partial x_i}{\partial q_k} \frac{\partial \dot{q}_k}{\partial \dot{q}_j} = 0 + \sum_k \frac{\partial x_i}{\partial q_k} \delta_{ij}$$

$$= \frac{\partial x_i}{\partial q_j}.$$

Thus, we conclude that

$$\frac{\partial v_i}{\partial \dot{q}_j} = \frac{\partial x_i}{\partial q_j}. \qquad (1.7)$$

This simple relationship comes in very handy in many derivations concerning generalized coordinates. Fortunately, it is easy to remember because it states that *the Cartesian velocity is related to the generalized velocity in the same way as the Cartesian coordinate is related to the generalized coordinate.*

1.4   Constraints

Every physical system has a particular number of *degrees of freedom*. The number of degrees of freedom is the number of independent coordinates needed to completely specify the position of every part of the system. To describe the position of a free particle one must specify the values of three coordinates (say, x, y and z). Thus a free particle has three degrees of freedom. For a system of two free particles you need to specify the positions of both particles. Each particle has three degrees of freedom, so the system as a whole has six degrees of freedom. In general a mechanical system consisting of N free particles will have $3N$ degrees of freedom.

When a system is acted upon by forces of constraint, it is often possible to reduce the number of coordinates required to describe the motion. Thus, for example, the motion of a particle on a table can be described in terms of x and y, the constraint being that the z coordinate (defined to be perpendicular to the table) is given by z = constant. The motion of a particle on the surface of a sphere can be described in terms of two angles, and the constraint itself can be expressed by r = constant. *Each constraint reduces by one the number of degrees of freedom.*

In many problems the system is constrained in some way. Examples are a marble rolling on the surface of a table or a bead sliding along a wire. In these problems the particle is not completely free. There are forces acting on it which restrict its motion. If a hockey puck is sliding on smooth ice, the gravitational force is acting downward and the normal force is acting upward. If the puck leaves the surface of the ice, the normal force ceases to act and the gravitational force quickly brings it back down to the surface. At the surface, the normal force prevents the puck from continuing to move downward in the vertical direction. The force exerted by the surface on the particle is called a *force of constraint*.

If a system of N particles is acted upon by k constraints, then the number of generalized coordinates needed to describe the motion is $3N - k$. (The number of Cartesian coordinates is always $3N$, but the number of (independent) generalized coordinates is $3N - k$.)

A constraint is a relationship between the coordinates. For example, if a particle is constrained to the surface of the paraboloid formed by rotating a parabola about the z axis, then the coordinates of the particle are related by $z - x^2/a - y^2/b = 0$. Similarly, a particle constrained to the surface of a sphere has coordinates that are related by $x^2 + y^2 + z^2 - a^2 = 0$.

If the equation of constraint (the relationship between the coordinates) can be expressed in the form

$$f(q_1, q_2, \ldots, q_n, t) = 0, \qquad (1.8)$$

then the constraint is called *holonomic*.[1]

The key elements in the definition of a holonomic constraint are: (1) the equals sign and (2) it is a relation involving the *coordinates*. For example, a possible constraint is that a particle is always outside of a sphere of radius a. This constraint could be expressed as $r \geq a$. This is *not* a holonomic constraint. Sometimes a constraint involves not just coordinates but also velocities or differentials of the coordinates. Such constraints are also not holonomic.

As an example of a non-holonomic constraint consider a marble rolling on a perfectly rough table. The marble requires five coordinates to completely describe its position and orientation, two linear coordinates to give its position on the table top and three angle coordinates to describe its orientation. If the table top is perfectly smooth, the marble can *slip* and there is no relationship between the linear and angular coordinates. If, however, the table top is rough there are relations (constraints) between the angle coordinates and the linear coordinates. These constraints will have the general form $dr = a d\theta$, which is a relation between differentials. If such differential expressions can be integrated, then the constraint becomes a relation between coordinates, and it is holonomic. But, in general, rolling on the surface of a plane does not lead to integrable relations. That is, in general, the *rolling* constraint is not holonomic because the rolling constraint is a relationship between *differentials*. The equation of constraint does not involve *only* coordinates. (But rolling in a straight line is integrable and hence holonomic.)

A holonomic constraint is an equation of the form of (1.8) relating the generalized coordinates, and it can be used to express one coordinate in terms of the others, thus reducing the number of coordinates required to describe the motion.

Exercise 1.5 A bead slides on a wire that is moving through space in a complicated way. Is the constraint holonomic? Is it scleronomous?

---

[1] A constraint that does not contain the time explicitly is called *scleronomous*. Thus $x^2 + y^2 + z^2 - a^2 = 0$ is both holonomic and scleronomous.

Exercise 1.6 Draw a picture showing that a marble can roll on a tabletop and return to its initial position with a different orientation, but that there is a one-to-one relation between angle and position for rolling in a straight line.

Exercise 1.7 Express the constraint for a particle moving on the surface of an ellipsoid.

Exercise 1.8 Consider a diatomic molecule. Assume the atoms are point masses. The molecule can rotate and it can vibrate along the line joining the atoms. How many rotation axes does it have? (We do not include operations in which the final state cannot be distinguished from the initial state.) How many degrees of freedom does the molecule have? (Answer: 6.)

Exercise 1.9 At room temperature the vibrational degrees of freedom of O_2 are "frozen out." How many degrees of freedom does the oxygen molecule have at room temperature? (Answer: 5.)

1.5 Virtual displacements

A *virtual displacement*, δx_i, is defined as an infinitesimal, instantaneous displacement of the coordinate x_i, consistent with any constraints acting on the system.

To appreciate the difference between an ordinary displacement dx_i and a virtual displacement δx_i, consider the transformation equations

$$x_i = x_i(q_1, q_2, \ldots, q_n, t) \quad i = 1, 2, \ldots, 3N.$$

Taking the differential of the transformation equations we obtain

$$dx_i = \frac{\partial x_i}{\partial q_1}dq_1 + \frac{\partial x_i}{\partial q_2}dq_2 + \cdots + \frac{\partial x_i}{\partial q_n}dq_n + \frac{\partial x_i}{\partial t}dt$$

$$= \sum_{\alpha=1}^{n} \frac{\partial x_i}{\partial q_\alpha}dq_\alpha + \frac{\partial x_i}{\partial t}dt.$$

But for a virtual displacement δx_i, time is frozen, so

$$\delta x_i = \sum_{\alpha=1}^{n} \frac{\partial x_i}{\partial q_\alpha}dq_\alpha. \tag{1.9}$$

1.6 Virtual work and generalized force

Suppose a system is subjected to a number of applied forces. Let F_i be the force component along x_i. (Thus, for a two-particle system, F_5 would be the force in the y direction acting on particle 2.) Allowing all the Cartesian coordinates to undergo virtual displacements δx_i, the *virtual work* performed by the applied forces will be

$$\delta W = \sum_{i=1}^{3N} F_i \delta x_i.$$

Substituting for δx_i from Equation (1.9)

$$\delta W = \sum_{i=1}^{3N} F_i \left(\sum_{\alpha=1}^{3N} \frac{\partial x_i}{\partial q_\alpha} \delta q_\alpha \right).$$

Interchanging the order of the summations

$$\delta W = \sum_{\alpha=1}^{3N} \left(\sum_{i=1}^{3N} F_i \frac{\partial x_i}{\partial q_\alpha} \right) \delta q_\alpha.$$

Harkening back to the elementary concept of "work equals force times distance" we define the *generalized force* as

$$Q_\alpha = \sum_{i=1}^{3N} F_i \frac{\partial x_i}{\partial q_\alpha}. \tag{1.10}$$

Then the virtual work can be expressed as

$$\delta W = \sum_{\alpha=1}^{3N} Q_\alpha \delta q_\alpha.$$

Equation (1.10) is the definition of generalized force.

Example 1.1 *Show that the principle of virtual work ($\delta W = 0$ in equilibrium) implies that the sum of the torques acting on the body must be zero.*

Solution 1.1 *Select an axis of rotation in an arbitrary direction $\boldsymbol{\Omega}$. Let the body rotate about $\boldsymbol{\Omega}$ through an infinitesimal angle ϵ. The virtual displacement of a point P_i located at $\mathbf{R}_i$ relative to the axis is given by*

$$\delta \mathbf{R}_i = \epsilon \boldsymbol{\Omega} \times \mathbf{R}_i.$$

The work done by a force $\mathbf{F}_i$ acting on P_i is

$$\delta W_i = \mathbf{F}_i \cdot \delta \mathbf{R}_i = \mathbf{F}_i \cdot (\epsilon \boldsymbol{\Omega} \times \mathbf{R}_i) = \epsilon \boldsymbol{\Omega} \cdot (\mathbf{R}_i \times \mathbf{F}_i) = \epsilon \boldsymbol{\Omega} \cdot \mathbf{N}_i,$$

where $\mathbf{N}_i$ *is the torque about the axis acting on* P_i. *The total virtual work is*

$$\delta W = \sum_i \epsilon \mathbf{\Omega} \cdot \mathbf{N}_i = \epsilon \mathbf{\Omega} \cdot \sum_i \mathbf{N}_i = \epsilon \mathbf{\Omega} \cdot \mathbf{N}_{tot}.$$

For a non-rotating body, the direction of $\mathbf{\Omega}$ *is arbitrary, so* $\mathbf{N}_{tot} = 0$.

Exercise 1.10 Consider a system made up of four particles described by Cartesian coordinates. What is F_{10}?

Exercise 1.11 A particle is acted upon by a force with components F_x and F_y. Determine the generalized forces in polar coordinates. Answer: $Q_r = F_x \cos\theta + F_y \sin\theta$ and $Q_\theta = -F_x r \sin\theta + F_y r \cos\theta$.

Exercise 1.12 Consider the arrangement in Figure 1.2. The string is inextensible and there is no friction. Show by evaluating the virtual work that equilibrium requires that $M = m/\sin\theta$. (This can be done trivially using elementary methods, but you are asked to solve the problem using virtual work.)

1.7   Configuration space

The specification of the positions of all of the particles in a system is called the *configuration* of the system. In general, the configuration of a system consisting of N free particles is given by

$$(q_1, q_2, \ldots, q_n), \quad \text{where } n = 3N. \tag{1.11}$$

Consider a single particle. Its position in Cartesian coordinates is given by x, y, z, and in a rectilinear three-dimensional reference frame the position of the particle is represented by a point. For a system consisting of *two* particles, you might represent the configuration of the system by two points. However, you could (at least conceptually) represent the configuration by a *single* point in a six-dimensional reference frame. In this 6-D space there would be six mutually perpendicular axes labeled, for example, $x_1, y_1, z_1, x_2, y_2, z_2$.

Similarly, **the configuration of a system composed of N particles is represented by a single point in a $3N$-dimensional reference frame.**

Figure 1.2   Two blocks in equilibrium.

The axes of such a multi-dimensional reference frame are not necessarily Cartesian. The positions of the various particles could be represented in spherical coordinates, or cylindrical coordinates, or the generalized coordinates, $q_1, q_2, \ldots, q_n$. Thus, the configuration of a system consisting of two particles is expressed as $(q_1, q_2, \ldots, q_6)$ where $q_1 = x_1, q_2 = y_1, \ldots, q_6 = z_2$ in Cartesian coordinates. In spherical coordinates, $q_1 = r_1, q_2 = \theta_1, \ldots, q_6 = \phi_2$, and similarly for any other coordinate system.

Consider the three-dimensional reference frame of a single particle. As time goes on the values of the coordinates change. This change is continuous and the point that represents the *position* of the particle will trace out a smooth curve. Similarly, in the n-dimensional configuration space with axes representing the generalized coordinates, the point representing the *configuration* of the system will trace out a smooth curve which represents the time development of the entire system. See Figure 1.3.

It might be noted that when we transform from one set of coordinates to a different set of coordinates, the character of the configuration space may change dramatically. Straight lines may become curves and angles and distances may change. Nevertheless, some geometrical properties remain unaltered. A point is still a point and a curve is still a curve. Adjacent curves remain adjacent and the neighborhood of a point remains the neighborhood of the point (although the shape of the neighborhood will probably be different).

Exercise 1.13 A particle moves along a line of constant r from θ_1 to θ_2. Show that this is a straight line in r–θ space but not in x–y space.

Exercise 1.14 The equation $y = 3x + 2$ describes a straight line in 2-D Cartesian space. Determine the shape of this curve in r–θ space.

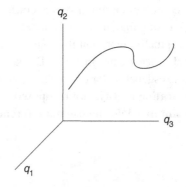

Figure 1.3   A path in configuration space. Note that time is a parameter giving the position of the configuration point along the curve.

1.8 Phase space

It is often convenient to describe a system of particles in terms of their positions and their momenta. The "state" of a system is described by giving the values of the positions and momenta of all the particles at some instant of time. For example we could represent the position of a particle by the Cartesian coordinates x_i, $i = 1, 2, 3$. Similarly, the momentum of the particle can be represented by p_i, $i = 1, 2, 3$. Imagine drawing a $2n$-dimensional coordinate system in which the axes are labeled thus: $x_1, \ldots, x_n$; $p_1, \ldots, p_n$. The space described by this set of axes is called *phase space*. The positions of *all* the particles as well as the momenta of *all* the particles are represented by a *single point* in phase space. As time progresses the position and the momentum of each particle changes. These change *continuously* so the point in phase space moves continuously in the $2n$ coordinate system. That is, the time development of the system can be represented by a *trajectory* in phase space. This trajectory is called the "phase path" or, in some contexts, the "world line."[2] We will shortly define "generalized momentum" and then phase space will be described in terms of the generalized coordinates $(q_1, \ldots, q_n)$ and the generalized momenta $(p_1, \ldots, p_n)$.

1.9 Dynamics

1.9.1 Newton's laws of motion

Dynamics is the study of the laws that determine the motion of material bodies. Your first introduction to dynamics was almost certainly a study of Isaac Newton's three laws.

(1) **The first law or the law of inertia**: a body not subjected to any external force will move in a straight line at constant speed.
(2) **The second law or the equation of motion**: the rate of change of momentum of a body is equal to the net external force applied to it.
(3) **The third law or action equals reaction**: if a body exerts a force on a second body, the second body exerts and equal and opposite force of the same kind on the first one.

The first law tells us that a *free particle* will move at constant velocity.

---

[2] The concept of phase space is very important in the study of chaotic systems where the intersection of the trajectory of the system with a particular plane in phase space (the so-called "surface of section") can be analyzed to determine if chaotic motion is taking place. Also in statistical mechanics, a basic theorem due to Liouville states that, if one plots the phase space trajectories of an *ensemble* of systems, the density of phase space points in the vicinity of a given system will remain constant in time.

The second law is usually expressed by the vector relation

$$\mathbf{F} = \frac{d\mathbf{p}}{dt}. \tag{1.12}$$

If the mass is constant this relationship reduces to the well known equation $\mathbf{F} = m\mathbf{a}$.

The third law can be expressed in either in the *strong form* or in the *weak form*. The strong form states that the forces are equal and opposite and directed along the line joining the particles. The weak form of the third law only states that the forces are equal and opposite.

These laws (especially the second law) are extremely useful for solving practical problems.

1.9.2 The equation of motion

The second law is often used to determine the acceleration of a body subjected to a variety of unbalanced forces. The equation for the acceleration in terms of positions and velocities is called the "equation of motion." (In introductory physics courses the determination of the acceleration involved isolating the system, drawing a free body diagram, and applying the second law, usually in the form $F = ma$.) Since forces can be expressed in terms of positions, velocities and time, the second law yields an expression for the acceleration having a form such as

$$\ddot{x} = \ddot{x}(x, \dot{x}, t).$$

This is the equation of motion. The position of a particle as a function of time can then be determined by integrating the equation of motion twice, generating

$$x = x(t).$$

The procedure is called *determining the motion*.

Newton's laws are simple and intuitive and form the basis for most introductory courses in mechanics. It will be convenient to occasionally refer to them, but our study is not based on Newton's laws.

1.9.3 Newton and Leibniz

It is well known that Isaac Newton and Gottfried Leibniz both invented the calculus independently. It is less well known that they had different notions concerning the time development of a system of particles. Newton's second law gives us a vector relationship between the force on a particle and its acceleration. (For a system of N particles, we have N vector relations corresponding to $3N$ scalar second-order differential equations. These equations are often coupled in the sense that the positions of all N particles may be present in each of the $3N$ equations of motion. To determine the motion we must solve all these coupled equations simultaneously.)

On the other hand, Leibniz believed that the motions of particles could be better analyzed by considering their *vis-viva* or (as we would call it today) their kinetic energy.

Essentially, Newton considered that the quantity $\Sigma m_i \mathbf{v}_i$ was conserved during the motion, whereas Leibniz believed that $\Sigma m_i v_i^2$ was constant. In modern terminology, Newton believed in the conservation of momentum whereas Leibniz believed in the conservation of kinetic energy. It soon became clear that kinetic energy (T) is not conserved during the motion of a system, but when the concept of potential energy (V) was introduced, Leibniz's theory was extended to state that the total energy $(E = T + V)$ is constant, which is, of course, one of the basic conservation laws of mechanics. Thus, you might say, both Newton and Leibniz were correct.

In your introductory physics course you often switched between using Newton's laws and using the conservation of energy. For example, the problem of a falling object can be solved using $F = ma$ or by applying the conservation of energy.

Later, Euler and Lagrange elaborated on Leibniz's idea and showed that the motion of a system can be predicted from a single unifying principle, which we now call Hamilton's principle. Interestingly, this approach does not involve vectors, or even forces, although forces can be obtained from it. The approach of Euler and Lagrange is referred to as "analytical mechanics" and is the primary focus of our study. (Since Newton's approach is very familiar to you, we shall occasionally use it when considering some aspect of a physical problem.) Analytical mechanics does not require Newton's three laws and, in particular, does away with the third law. Some physicists have suggested that Newton introduced the third law as a way of dealing with constraints. (A bouncing ball is "constrained" to remain inside the room. In the Newtonian picture, the ball bounces off a wall because the wall exerts a reaction force on it, as given by the third law.) The Lagrangian method incorporates constraints into the framework of the analysis, and neither the second nor the third law is necessary. However, the forces of constraint as well as the equation of motion can be determined from the method. Furthermore, all of the equations in analytical mechanics are *scalar* equations, so we do not need to use any concepts from vector analysis. However, sometimes vectors are convenient and we shall use them occasionally.

This book is concerned with developing the concepts and techniques of analytical mechanics using variational principles.[3] I am sure you will find this to be a very beautiful and very powerful theory. In the introduction to a well known book on the subject,[4] the author, speaking about himself, says, "Again and again he experienced the extraordinary elation of mind which accompanies a preoccupation with the basic principles and methods of analytical mechanics." (You may not feel an "elation of mind", but I think you will appreciate what the author was expressing.)

1.10 Obtaining the equation of motion

In this section we review the methods for obtaining and solving the equation of motion. As you might expect, solving the equation of motion yields an expression for the *motion*, that is, for the position as a function of time. It happens that there are several different ways of expressing the equation of motion, but for now let us think of it as an expression for the acceleration of a particle in terms of the positions and velocities of all the other particles comprising the system.

As pointed out by Landau and Lifshitz,[5] "If all the coordinates and velocities are simultaneously specified, it is known from experience that the state of the system is completely determined and that its subsequent motion can, in principle, be calculated. Mathematically this means that if all the coordinates q and velocities $\dot{q}$ are given at some instant, the accelerations $\ddot{q}$ at that instant are uniquely defined."

In other words, the equations of motion can be expressed as relations of the type

$$\ddot{q}_i = \ddot{q}_i(q_1, q_2, \ldots, q_n; \dot{q}_1, \dot{q}_2, \ldots, \dot{q}_n; t).$$

If we know the accelerations ($\ddot{q}_i$) of the particles, then we can (in principle) determine the positions and velocities at a subsequent time. Thus, a knowledge of the equations of motion allows us to predict the time development of a system.

---

[3] Be aware that the fields of analytical mechanics and the calculus of variations are vast and this book is limited to presenting some fundamental concepts.

[4] Cornelius Lanczos, *The Variational Principles of Mechanics,* The University of Toronto Press, 1970. Reprinted by Dover Press, New York, 1986.

[5] L. D. Landau and E. M. Lifshitz, *Mechanics, Vol 1 of A Course of Theoretical Physics,* Pergamon Press, Oxford, 1976, p.1.

1.10.1 The equation of motion in Newtonian mechanics

If the masses are constant, an elementary way to obtain the equation of motion is to use Newton's second law in the form

$$\ddot{\mathbf{r}} = \mathbf{F}/m. \tag{1.13}$$

This is a second-order differential equation for $\mathbf{r}$ that can, in principle, be solved to yield $\mathbf{r} = \mathbf{r}(t)$. Of course, this requires having an expression for the force as a function of $\mathbf{r}$, $\dot{\mathbf{r}}$, and t.

As a simple one-dimensional example, consider a mass m connected to a spring of constant k. The force exerted by the spring on the mass is $F = -kx$. Therefore, Newton's second law yields the following equation of motion

$$m\ddot{x} + kx = 0. \tag{1.14}$$

It is important to note that the second law is applicable only in inertial coordinate systems. Newton was aware of this problem and he stated that the second law is a relationship that holds in a coordinate system *at rest with respect to the fixed stars.*

1.10.2 The equation of motion in Lagrangian mechanics

Another way to obtain the equation of motion is to use the Lagrangian technique.

When dealing with particles or with rigid bodies that can be treated as particles, the *Lagrangian* can be defined to be the difference between the kinetic energy and the potential energy.[6] That is,

$$L = T - V. \tag{1.15}$$

For example, if a mass m is connected to a spring of constant k, the potential energy is $V = \frac{1}{2}kx^2$ and the kinetic energy is $T = \frac{1}{2}mv^2 = \frac{1}{2}m\dot{x}^2$. Therefore the Lagrangian is

$$L = T - V = \frac{1}{2}m\dot{x}^2 - \frac{1}{2}kx^2.$$

It is usually easy to express the potential energy in whatever set of coordinates being used, but the expression for the kinetic energy may be somewhat difficult to determine, so whenever possible, one should start with Cartesian

---

[6] Although Equation (1.15) is pefectly correct for a system of particles, we shall obtain a somewhat different expression when considering continuous systems in Chapter 7. In general, the Lagrangian is defined to be a function that generates the equations of motion.

coordinates. In the Cartesian coordinate system the kinetic energy takes on the particularly simple form of a sum of the velocities squared. That is,

$$T = \frac{1}{2}m(\dot{x}^2 + \dot{y}^2 + \dot{z}^2).$$

To express the kinetic energy in terms of some other coordinate system requires a set of transformation equations.

For example, for a pendulum of length l, the potential energy is $V = -mgl \cos\theta$ and the kinetic energy is $T = \frac{1}{2}mv^2 = \frac{1}{2}m(l\dot{\theta})^2$. (Here θ is the angle between the string and the perpendicular.) Therefore the Lagrangian is

$$L = T - V = \frac{1}{2}ml^2\dot{\theta}^2 + mgl \cos\theta.$$

The kinetic energy is a function of the velocity. (You are familiar with expressions such as $T = \frac{1}{2}m\dot{x}^2$ and $T = \frac{1}{2}ml^2\dot{\theta}^2$.) The velocity is the time derivative of a position coordinate ($\dot{x}$ and $l\dot{\theta}$). The potential energy is usually a function only of the position. In generalized coordinates, the position is expressed as q and the velocity as $\dot{q}$. The Lagrangian is, therefore, a function of q and $\dot{q}$. That is, $L = L(q, \dot{q})$, or more generally, $L = L(q, \dot{q}, t)$. If we are concerned with a single particle free to move in three dimensions, there are three qs and three $\dot{q}$s. Then $L = L(q_1, q_2, q_3, \dot{q}_1, \dot{q}_2, \dot{q}_3, t)$. For a system of N particles, if $n = 3N$ we write

$$L = L(q_i, \dot{q}_i, t); \qquad i = 1, \ldots, n.$$

I assume you were introduced to the Lagrangian and Lagrange's equations in your course on intermediate mechanics. In the next two chapters you will find derivations of Lagrange's equations from first principles. However, for the moment, I will simply express the equation without proof and show you how to use it as a tool for obtaining the equation of motion.

For a single coordinate q, Lagrange's equation is

$$\frac{d}{dt}\frac{\partial L}{\partial \dot{q}} - \frac{\partial L}{\partial q} = 0. \tag{1.16}$$

If there are n coordinates, there are n Lagrange equations, namely,

$$\frac{d}{dt}\frac{\partial L}{\partial \dot{q}_i} - \frac{\partial L}{\partial q_i} = 0, \quad i = 1, \ldots, n. \tag{1.17}$$

It is important to realize that **Lagrange's equations are the equations of motion of a system.**

For example, for a mass on a spring the Lagrangian is $L = T - V = \frac{1}{2}m\dot{x}^2 - \frac{1}{2}kx^2$. Plugging this into the Lagrange equation we obtain

$$\frac{d}{dt}\left(\frac{\partial L}{\partial \dot{x}}\right) - \frac{\partial L}{\partial x} = 0,$$

$$\frac{d}{dt}\frac{\partial}{\partial \dot{x}}\left(\frac{1}{2}m\dot{x}^2 - \frac{1}{2}kx^2\right) - \frac{\partial}{\partial x}\left(\frac{1}{2}m\dot{x}^2 - \frac{1}{2}kx^2\right) = 0,$$

$$\frac{d}{dt}(m\dot{x}) + kx = 0,$$

so

$$m\ddot{x} + kx = 0,$$

as expected. (Compare with Equation 1.14.)

To illustrate the use of Lagrange's equations to obtain the equations of motion, we consider several simple mechanical systems.

Example 1.2 *Atwood's machine: Figure 1.4 is a sketch of Atwood's machine. It consists of masses m_1 and m_2 suspended by a massless inextensible string over a frictionless, massless pulley. Evaluate the Lagrangian and obtain the equation of motion.*

Solution 1.2 *The kinetic energy of the masses is*

$$T = \frac{1}{2}m_1\dot{x}_1^2 + \frac{1}{2}m_2\dot{x}_2^2,$$

and the potential energy is

$$V = -m_1gx_1 - m_2gx_2,$$

where we selected $V = 0$ at the center of the pulley. The system is subjected to the constraint $x_1 + x_2 = l = $ constant. The Lagrangian is,

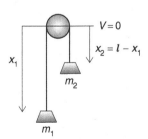

Figure 1.4   Atwood's machine.

$$L = T - V = \frac{1}{2}m_1\dot{x}_1^2 + \frac{1}{2}m_2\dot{x}_2^2 + m_1gx_1 + m_2gx_2.$$

But using $x_2 = l - x_1$ we can rewrite the Lagrangian in terms of a single variable,

$$L = \frac{1}{2}m_1\dot{x}_1^2 + \frac{1}{2}m_2\dot{x}_1^2 + m_1gx_1 + m_2g(l - x_1),$$

$$= \frac{1}{2}(m_1 + m_2)\dot{x}_1^2 + (m_1 - m_2)gx_1 + m_2gl.$$

The equation of motion is

$$\frac{d}{dt}\frac{\partial L}{\partial \dot{x}_1} - \frac{\partial L}{\partial x_1} = 0,$$

$$\frac{d}{dt}(m_1 + m_2)\dot{x}_1 - (m_1 - m_2)g = 0,$$

$$\ddot{x}_1 = \frac{m_1 - m_2}{m_1 + m_2}g.$$

Example 1.3 *A cylinder of radius a and mass m rolls without slipping on a fixed cylinder of radius b. Evaluate the Lagrangian and obtain the equation of motion for the short period of time before the cylinders separate. See Figure 1.5*

Solution 1.3 *The kinetic energy of the smaller cylinder is the kinetic energy of translation of its center of mass plus the kinetic energy of rotation about its center of mass. That is,*

$$T = \frac{1}{2}m\left(\dot{r}^2 + r^2\dot{\theta}^2\right) + \frac{1}{2}I\dot{\phi}^2,$$

where r is the distance between the centers of the cylinders and $I = \frac{1}{2}ma^2$. The potential energy is $V = mgr\cos\theta$. Therefore

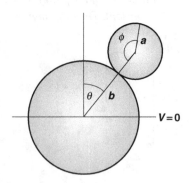

Figure 1.5   A cylinder rolling on another cylinder.

$$L = \frac{1}{2}m(\dot{r}^2 + r^2\dot{\theta}^2) + \frac{1}{2}I\dot{\phi}^2 - mgr\cos\theta.$$

But there are two constraints, namely, $r = a + b$ and $a\phi = b\theta$. Therefore $\dot{r} = 0$ and $\dot{\phi} = (b/a)\dot{\theta}$. So,

$$L = \frac{1}{2}m(a+b)^2\dot{\theta}^2 + \frac{1}{2}\left(\frac{1}{2}ma^2\right)\left(\frac{b}{a}\dot{\theta}\right)^2 - mg(a+b)\cos\theta$$

$$= \frac{1}{2}m\left(a^2 + 2ab + \frac{3}{2}b^2\right)\dot{\theta}^2 - mg(a+b)\cos\theta$$

and the equation of motion is

$$\frac{d}{dt}\frac{\partial L}{\partial \dot{q}} - \frac{\partial L}{\partial q} = 0$$

or

$$m\left(a^2 + 2ab + \frac{3}{2}b^2\right)\ddot{\theta} - mg(a+b)\sin\theta = 0.$$

Example 1.4 *A bead of mass m slides on a massive frictionless hoop of radius a. The hoop is rotating at a constant angular speed ω about a vertical axis. See Figure 1.6. Determine the Lagrangian and the equation of motion.*

Solution 1.4 *If the hoop is massive enough, the motion of the bead will not affect the rotation rate of the hoop. Thus we ignore the energy of the hoop. The kinetic energy is the energy of the bead due to the rotation of the hoop as well as kinetic energy of the bead due to its motion on the hoop which is associated with changes in the angle θ. Therefore,*

$$L = \frac{1}{2}ma^2\dot{\theta}^2 + \frac{1}{2}ma^2\omega^2\sin^2\theta + mga\cos\theta.$$

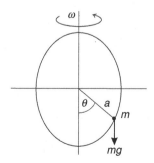

Figure 1.6   A bead slides on a rotating hoop.

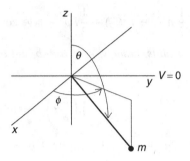

Figure 1.7   A spherical pendulum.

The equation of motion is

$$ma^2\ddot{\theta} - ma^2\omega^2 \sin\theta\cos\theta + mga\sin\theta = 0.$$

Example 1.5 *A spherical pendulum consists of a bob of mass m suspended from an inextensible string of length l. Determine the Lagrangian and the equations of motion. See Figure 1.7. Note that θ, the polar angle, is measured from the positive z axis, and φ, the azimuthal angle, is measured from the x axis in the x–y plane.*

Solution 1.5 *The Lagrangian is*

$$L = \frac{1}{2}m(\dot{x}^2 + \dot{y}^2 + \dot{z}^2) - mgz.$$

But

$$x = l\sin\theta\cos\phi$$
$$y = l\sin\theta\sin\phi$$
$$z = l\cos\theta$$

so

$$L = \frac{1}{2}ml^2(\dot{\theta}^2 + \sin^2\theta\dot{\phi}^2) - mgl\cos\theta.$$

The equations of motion are

$$\frac{d}{dt}\left(ml^2\dot{\theta}\right) - ml^2\dot{\phi}^2\sin\theta\cos\theta - mgl\sin\theta = 0,$$

$$\frac{d}{dt}(ml^2\sin^2\theta\dot{\phi}) = 0.$$

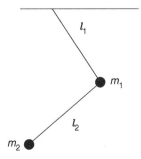

Figure 1.8   A double planar pendulum.

Exercise 1.15 Obtain the Lagrangian and the equation of motion for an Atwood's machine if the pulley has a moment of inertia I. Answer: $a = g(m_2 - m_1)/(m_1 + m_2 + I/R^2)$.

Exercise 1.16 Obtain the Lagrangian for a simple pendulum and for a double pendulum. (The double pendulum is illustrated in Figure 1.8. Assume the motion is planar.) Answer: For the double pendulum

$$L = \frac{1}{2}(m_1 + m_2)l_1^2\dot{\theta}_1^2 + \frac{1}{2}m_2\left(l_2^2\dot{\theta}_2^2 + 2l_1l_2\dot{\theta}_1\dot{\theta}_2\cos(\theta_1 - \theta_2)\right)$$
$$+ m_1gl_1\cos\theta_1 + m_2g(l_1\cos\theta_1 + l_2\cos\theta_2).$$

Exercise 1.17 A flyball governor is a simple mechanical device to control the speed of a motor. As the device spins faster and faster, the mass m_2 rises. The rising mass controls the fuel supply to the motor. See Figure 1.9. The angle θ is formed by one of the rods and the central axis, and the angle ϕ is the rotation angle about the axis. The angular speed at any moment is $\omega = \dot{\phi}$. Assume the system is in a uniform gravitational field. Show that the Lagrangian is given by

$$L = m_1a^2\left(\dot{\theta}^2 + \omega^2\sin^2\theta\right) + 2m_2a^2\dot{\theta}^2\sin^2\theta + 2(m_1 + m_2)ga\cos\theta.$$

1.11   Conservation laws and symmetry principles

The three most important conservation laws in classical mechanics are the conservation of linear momentum, the conservation of angular momentum and the conservation of energy. Although you are very familiar with these laws,

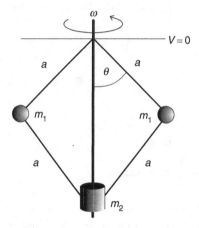

Figure 1.9   A flyball governor.

it is interesting to consider them in terms of generalized coordinates and the Lagrangian. We will show that these three basic conservation laws are consequences of the homogeneity and isotropy of space and the homogeneity of time.

When we say a region of space is *homogeneous* we mean that it is the same in one location as in another. Consequently, a system will be unaffected by a displacement from one point to another. As a simple example, I claim that my grandfather clock behaved the same on one side of the room as on the other (but it would behave differently on the Moon).

Similarly, if space is *isotropic,* it looks the same in all directions and a physical system is unchanged under a rotation. My grandfather clock behaved the same after I rotated it through 90° about a vertical axis. However, the space in my room is not isotropic for a rotation about a horizontal axis. (When I turned my grandfather clock upside-down it stopped working.)

Finally, if time is homogeneous, a physical system will behave the same at two different times. My clock behaved the same today as it did last week.

Time is also isotropic in the sense that physical systems behave the same whether time is running forward or backward.

When we say that a system has a particular symmetry, we mean that the system is invariant with respect to a change in some particular coordinate. For example, a sphere is an object that appears the same no matter how it is rotated. Spherical symmetry implies an invariance with respect to rotations, i.e., with respect to a change in the angular orientation of the object. Similarly, if a system is unchanged when it undergoes a displacement in some particular direction, we say that it has translational symmetry in that direction. When a system is invariant with respect to a change in time, we say it has temporal symmetry.

As an example of translational symmetry consider an object sitting on an infinite horizontal surface in a uniform, vertical gravitational field. If the object is moved in the horizontal plane, the system is unchanged. If, however, the object is moved vertically (if it is raised), the potential energy changes. Thus this system has translational symmetry in the horizontal plane but it does not have translational symmetry for vertical displacements. In a region where no forces are acting, space is homogeneous in all three directions.

A mechanical system is completely described by its Lagrangian. If the Lagrangian does not depend explicitly on some particular coordinate q_i, then a change in q_i does not affect the system. The system is said to be symmetric with respect to changes in q_i.

1.11.1 Generalized momentum and cyclic coordinates

Generalized momentum

A free particle is described by the Lagrangian

$$L = \frac{1}{2}m(\dot{x}^2 + \dot{y}^2 + \dot{z}^2).$$

Taking the partial derivative with respect to $\dot{x}$ we obtain

$$\frac{\partial L}{\partial \dot{x}} = m\dot{x}.$$

But $m\dot{x}$ is just the x component of the linear momentum. That is,

$$p_x = \frac{\partial L}{\partial \dot{x}}. \tag{1.18}$$

Similarly,

$$p_y = \frac{\partial L}{\partial \dot{y}} \qquad p_z = \frac{\partial L}{\partial \dot{z}}.$$

The Lagrangian for a freely rotating wheel with moment of inertia I is

$$L = \frac{1}{2}I\dot{\theta}^2$$

and

$$\frac{\partial L}{\partial \dot{\theta}} = I\dot{\theta}.$$

But $I\dot{\theta}$ is the angular momentum.

In both of these examples, the momentum was the derivative of the Lagrangian with respect to a velocity. We carry this idea to its logical conclusion and define the *generalized momentum* by

$$p_i = \frac{\partial L}{\partial \dot{q}_i}.$$

The generalized momentum p_i is associated with the generalized coordinate q_i and is sometimes referred to as the *conjugate momentum*. Thus, we have seen that the linear momentum p_x is conjugate to the linear coordinate x and the angular momentum $I\dot{\theta}$ is conjugate to the angular coordinate θ.

Cyclic coordinates

If a particular coordinate does not appear in the Lagrangian, it is called "*cyclic*" or "*ignorable*." For example, the Lagrangian for a mass point in a gravitational field is

$$L = \frac{1}{2}m(\dot{x}^2 + \dot{y}^2 + \dot{z}^2) - mgz.$$

Since neither x nor y appear in the Lagrangian, they are cyclic. (Note, however, that the time derivatives of the coordinates, $\dot{x}$ and $\dot{y}$, do appear in the Lagrangian, so there is an implicit dependence on x and y.)

Lagrange's equation tells us that

$$\frac{d}{dt}\frac{\partial L}{\partial \dot{q}_i} - \frac{\partial L}{\partial q_i} = 0.$$

But if q_i is cyclic, the second term is zero and

$$\frac{d}{dt}\frac{\partial L}{\partial \dot{q}_i} = 0.$$

Since

$$p_i = \frac{\partial L}{\partial \dot{q}_i}$$

we have

$$\frac{d}{dt}p_i = 0$$

or

$$p_i = \text{constant}.$$

That is, **the generalized momentum conjugate to a cyclic coordinate is a constant.**

Exercise 1.18 Identify any conserved momenta for the spherical pendulum.

Exercise 1.19 Write the Lagrangian for a planet orbiting the Sun. Determine any cyclic coordinates and identify the conserved conjugate momenta. Answer: $L = (1/2)(m\dot{r}^2 + mr^2\dot{\theta}^2) + GmM/r^2$.

Exercise 1.20 Show that as long as the kinetic energy depends only on velocities and not on coordinates, then the generalized force is given by $Q_i = \frac{\partial L}{\partial q_i}$.

The relation between symmetries and conserved quantities

We have determined that the generalized momentum conjugate to an ignorable (cyclic) coordinate is conserved. If the coordinate q_i is cyclic, it does not appear in the Lagrangian and the system does not depend on the value of q_i. Changing q_i has no effect on the system; it is the same after q_i is changed as it was before. If $q_i = \theta$, then a rotation through θ changes nothing. But that is, essentially, the definition of symmetry. Consequently, if a coordinate is cyclic, the system is symmetric under changes in that coordinate. Furthermore, each symmetry in a coordinate gives rise to a conserved conjugate momentum. For example, for an object sitting on a horizontal surface, the horizontal components of linear momentum are constant. If the x and y axes are in the horizontal plane, then there is no force in either the x or y directions. If there were a force in, say, the x direction, then the potential energy would depend on x and consequently the Lagrangian would contain x and it would not be an ignorable coordinate.

The idea that conserved quantities are related to symmetries is highly exploited in studies of the elementary particles. For example, particle physicists frequently utilize the law of conservation of parity and the law of conservation of "strangeness." These conservation laws are associated with observed symmetries in the behavior of elementary particles. Parity conservation is an expression of the symmetry of right-handedness and left-handedness in reactions involving elementary particles. (As you probably know, there are some reactions in which parity is not conserved.)

As an example from the field of elementary particle physics, it is observed that some reactions never occur. There is no apparent reason why a reaction such as $\pi^- + p \to \pi^o + \Lambda$ does not take place. It must violate some conservation law. However, it does not violate any of the everyday conservation laws such as charge, mass-energy, parity, etc. But since the reaction does not occur, using the principle that "what is not forbidden is required," physicists concluded that there must be a conservation principle at work. They called it conservation of strangeness. Studying reactions that do occur allowed them to assign to each particle a "strangeness quantum number." For example, the strangeness of a π particle is 0, that of a proton is also 0 and the strangeness of a Λ particle is -1. The total strangeness on the right-hand side of the reaction is not equal to the total strangeness on the left-hand side. Therefore, in this reaction, strangeness is not conserved and the reaction does not take place. This is not any weirder than requiring that linear momentum be conserved during a collision. However, we feel at home with momentum conservation because we have an analytical expression for momentum and we known that conservation of momentum implies that no net external forces are acting on the system. We do not have an analytical expression for strangeness nor do we know what symmetry is implied by strangeness conservation. We have no idea what the associated cyclic coordinate might be. None the less, the basic idea is the same.

In conclusion, we can state that *for every symmetry, there is a corresponding constant of the motion.* This is, essentially, Noether's theorem.[7]

Exercise 1.21 Find the ignorable coordinates for the spherical pendulum and determine the associated symmetry.

1.11.2 The conservation of linear momentum

In this section we will study the theoretical basis for the law of conservation of linear momentum. However, before we begin, let us note that there are several different ways to arrive at this conservation law.

For example, in your introductory physics course, you learned the law of conservation of linear momentum by considering Newton's second law in the form

$$\mathbf{F} = \frac{d\mathbf{P}}{dt}.$$

If there is no net external force acting on the system, $\mathbf{F} = 0$, and hence the time derivative of the total momentum $\mathbf{P}$ is zero, that is, the total linear momentum of the system is constant.

In your intermediate mechanics course, you probably considered the conservation of linear momentum from a more sophisticated point of view. Perhaps

---

[7] Emmy Noether (1882–1932), a mathematical physicist, proved the theorem that is named after her.

you considered a situation in which the potential energy did not depend on some particular coordinate, say x, and the kinetic energy also did not contain x. When you wrote the Lagrangian, $L = T - V$, you obtained an expression that did not contain x and consequently x was a cyclic coordinate. That is,

$$\frac{\partial L}{\partial x} = 0.$$

Now the Lagrange equation for x is

$$\frac{d}{dt}\frac{\partial L}{\partial \dot{x}} - \frac{\partial L}{\partial x} = 0,$$

which reduces to

$$\frac{d}{dt}\frac{\partial L}{\partial \dot{x}} = 0.$$

But $\frac{\partial L}{\partial \dot{x}} = p_x$, so

$$\frac{d}{dt}p_x = 0.$$

Again we see that if the Lagrangian does not depend on x, then p_x is constant.

We now show that the conservation of linear momentum is a consequence of the homogeneity of space, that is, we shall show that in a region of space whose properties are the same everywhere, the total linear momentum of a mechanical system will be constant. Our argument not only guarantees the conservation of linear momentum, but also leads to Newton's second and third laws, as we now demonstrate.

Consider a system composed of N particles situated at $\mathbf{r}_\alpha$ ($\alpha = 1, \ldots, N$). (For the sake of variety, the argument here is formulated in terms of vectors.) If every particle is displaced by the same infinitesimal distance $\boldsymbol{\epsilon}$, the positions of all particles will change according to $\mathbf{r}_\alpha \to \mathbf{r}_\alpha + \boldsymbol{\epsilon}$. We assume that $\boldsymbol{\epsilon}$ is a virtual displacement in which the all the particles are displaced infinitesimally but the velocities are unchanged and time is frozen. Recall that if $f = f(q_i, \dot{q}_i, t)$, then the differential of f, according to the rules of calculus, is

$$df = \sum_{i=1}^{n} \frac{\partial f}{\partial q_i} dq_i + \sum_{i=1}^{n} \frac{\partial f}{\partial \dot{q}_i} d\dot{q}_i + \frac{\partial f}{\partial t} dt.$$

But the change in f due to an infinitesimal virtual displacement is

$$\delta f = \frac{\partial f}{\partial q_1}\delta q_1 + \frac{\partial f}{\partial q_2}\delta q_2 + \cdots$$

$$= \sum_{\alpha=1}^{N} \frac{\partial f}{\partial \mathbf{r}_a} \cdot \delta \mathbf{r}_\alpha,$$

where we have used the fact that time is frozen and that a displacement of the entire system will have no effect on velocities.[8]

Consequently, the change in a Lagrangian due to a virtual displacement ϵ is

$$\delta L = \sum_\alpha \frac{\partial L}{\partial \mathbf{r}_\alpha} \cdot \delta \mathbf{r}_\alpha = \sum_\alpha \frac{\partial L}{\partial \mathbf{r}_\alpha} \cdot \epsilon = \epsilon \cdot \sum_\alpha \frac{\partial L}{\partial \mathbf{r}_\alpha},$$

where we used that fact that $\delta \mathbf{r}_\alpha = \epsilon$, that is, all particles are displaced the same amount. Note that we are summing over particles ($\alpha = 1, \ldots, N$), not components ($i = 1, \ldots, 3N$).

If the space is homogeneous, the Lagrangian does not change and $\delta L = 0$. Therefore,

$$\sum_\alpha \frac{\partial L}{\partial \mathbf{r}_\alpha} = 0.$$

But by Lagrange's equations,

$$\frac{\partial L}{\partial \mathbf{r}_\alpha} = \frac{d}{dt} \frac{\partial L}{\partial \mathbf{v}_\alpha},$$

so

$$\frac{d}{dt} \sum_\alpha \frac{\partial L}{\partial \mathbf{v}_\alpha} = \frac{d}{dt} \sum_\alpha \mathbf{p}_\alpha = \frac{d}{dt} \mathbf{P}_{\text{tot}} = 0.$$

This means that

$$\mathbf{P}_{\text{tot}} = \text{constant},$$

as expected.

We are not using any particular set of coordinates, as we have been formulating the problem in terms of vectors. The vector definition of kinetic energy is $T = (1/2)m\mathbf{v} \cdot \mathbf{v}$. As long as we express our relations in terms of vectors

---

[8] We are using the notation of Landau and Lifshitz involving the derivative of a scalar with respect to a vector. For example, if there is only one particle ($\alpha = 1$), we define

$$\frac{\partial F}{\partial \mathbf{r}} \equiv \hat{i} \frac{\partial F}{\partial x} + \hat{j} \frac{\partial F}{\partial y} + \hat{k} \frac{\partial F}{\partial z}.$$

Thus, the derivative of a scalar with respect to a vector is defined as the vector whose components are the derivatives of the scalar with respect to the components of the vector. Note that

$$\frac{\partial F}{\partial \mathbf{r}} = \nabla F.$$

the kinetic energy depends only on the square of the velocity and not on the coordinates[9] and we can write

$$\frac{\partial L}{\partial \mathbf{r}_\alpha} = -\frac{\partial V}{\partial \mathbf{r}_\alpha} = \mathbf{F}_\alpha.$$

But if

$$\sum_\alpha \frac{\partial L}{\partial \mathbf{r}_\alpha} = 0,$$

then

$$\sum_\alpha \mathbf{F}_\alpha = 0.$$

That is, the sum of all the forces acting on all the particles in the system is zero. If the system consists of just two particles, this tells us that $\mathbf{F}_1 + \mathbf{F}_2 = 0$, which is Newton's third law. Additionally,

$$\frac{\partial L}{\partial \mathbf{r}_\alpha} = \frac{d}{dt}\frac{\partial L}{\partial \mathbf{v}_\alpha} = \frac{d}{dt}\mathbf{p}_\alpha = \mathbf{F}_\alpha,$$

and we have also obtained Newton's second law.

1.11.3 The conservation of angular momentum

Our next task is to show that the conservation of angular momentum is a consequence of the isotropy of space. But before we tackle that problem, let us consider the conservation law first from an elementary and then from an intermediate point of view.

The *elementary level* consideration of the law of conservation of angular momentum usually starts with the definition of the angular momentum of a particle:

$$\mathbf{l} = \mathbf{r} \times \mathbf{p},$$

where $\mathbf{r}$ is the position of the particle and $\mathbf{p} = m\mathbf{v}$ is its linear momentum. Differentiating with respect to time yields

$$\frac{d\mathbf{l}}{dt} = \frac{d}{dt}(\mathbf{r} \times \mathbf{p}) = \frac{d\mathbf{r}}{dt} \times \mathbf{p} + \mathbf{r} \times \frac{d\mathbf{p}}{dt}.$$

---

[9] This is also true in Cartesian coordinates in which $T = (1/2)m\left(\dot{x}^2 + \dot{y}^2 + \dot{z}^2\right)$. But in most other coordinate systems the kinetic energy will depend on coordinates as well as velocities. For example, in spherical coordinates

$$T = (1/2)m(\dot{r}^2 + r^2\dot{\theta}^2 + r^2 \sin^2\theta\dot{\phi}^2).$$

Since $\frac{d\mathbf{r}}{dt} = \mathbf{v}$, the first term is $m(\mathbf{v} \times \mathbf{v}) = 0$. The second term contains $\frac{d\mathbf{p}}{dt}$ which, by Newton's second law, is the net force acting on the particle. Therefore the second term is $\mathbf{r} \times \mathbf{F} = \mathbf{N} = \text{torque}$. That is

$$\frac{d\mathbf{l}}{dt} = \mathbf{N}.$$

One can then show that the law holds for a system of particles (which requires invoking Newton's third law), and one concludes that the angular momentum of a system of particles is constant if there is no net external external torque acting on it.

A familiar example is the constancy of angular momentum of a particle in a central force field (such as a planet under the influence of the Sun). Since a central force has the form $\mathbf{F} = f(r)\hat{\mathbf{r}}$, the torque acting on the particle is

$$\mathbf{N} = \mathbf{r} \times \mathbf{F} = rf(r)(\hat{\mathbf{r}} \times \hat{\mathbf{r}}) = 0,$$

and the conjecture is proved.

An *intermediate level* discussion of the conservation of angular momentum might begin by considering a particle moving in a plane in a region of space where the potential energy is $V = V(r)$. (For example, $V = -GMm/r$.) The transformation equations from Cartesian to polar coordinates are

$$x = r \cos \theta$$
$$y = r \sin \theta.$$

Therefore,

$$\dot{x} = \dot{r} \cos \theta - r\dot{\theta} \sin \theta$$
$$\dot{y} = \dot{r} \sin \theta + r\dot{\theta} \cos \theta,$$

and

$$T = \frac{1}{2}m\left(\dot{x}^2 + \dot{y}^2\right) = \frac{1}{2}m\left(\dot{r}^2 + r^2\dot{\theta}^2\right)$$

so

$$L = T - V = \frac{1}{2}m\left(\dot{r}^2 + r^2\dot{\theta}^2\right) - V(r).$$

The momentum conjugate to θ is

$$p_\theta = \frac{\partial L}{\partial \dot{\theta}} = \frac{\partial}{\partial \dot{\theta}}\left[\frac{1}{2}m\left(\dot{r}^2 + r^2\dot{\theta}^2\right) - V(r)\right] = mr^2\dot{\theta},$$

which we recognize as the angular momentum of the particle. Observe that θ is ignorable, so the angular momentum, $mr^2\dot{\theta}$ is constant.

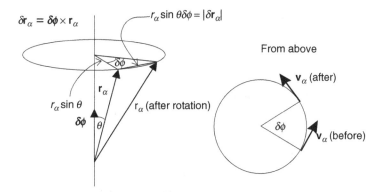

Figure 1.10   The rotation of the system changes the position of each particle by $\delta\mathbf{r}_\alpha$ and the velocity of each particle by $\delta\mathbf{v}_\alpha$.

We now address this question in a more *advanced* manner and show that the conservation of angular momentum is a consequence of the isotropy of space. Consider a mechanical system that is rotated through a virtual angle $\delta\phi$ around some axis. Let $\delta\boldsymbol{\phi}$ be a vector directed along the axis of rotation and having magnitude $\delta\phi$. Place the origin of coordinates on the axis of rotation. Owing to the rotation, every particle in the system is displaced by some distance $\delta\mathbf{r}_\alpha$ and if a particle has velocity $\mathbf{v}_\alpha$, the direction of the velocity vector will change by $\delta\mathbf{v}_\alpha$. Figure 1.10 shows the effect of the rotation on the position vector $\mathbf{r}_\alpha$.

If the space is isotropic, the rotation will have no effect on the Lagrangian so $\delta L = 0$. That is,

$$0 = \delta L = \sum_\alpha \left(\frac{\partial L}{\partial \mathbf{r}_\alpha} \cdot \delta\mathbf{r}_\alpha + \frac{\partial L}{\partial \mathbf{v}_\alpha} \cdot \delta\mathbf{v}_\alpha \right).$$

(Note that in this situation the velocities of the particles are not constant.) Again, we use Lagrange's equation to replace $\frac{\partial L}{\partial \mathbf{r}_\alpha}$ with $\frac{d}{dt}\frac{\partial L}{\partial \mathbf{v}_\alpha} = \frac{d}{dt}\mathbf{p}_\alpha = \dot{\mathbf{p}}_\alpha$ and write

$$\sum_\alpha \left(\dot{\mathbf{p}}_\alpha \cdot \delta\mathbf{r}_\alpha + \mathbf{p}_\alpha \cdot \delta\mathbf{v}_\alpha \right) = 0.$$

From the figure, $|\delta\mathbf{r}| = r\sin\theta\,\delta\phi$. But since $\delta\mathbf{r}$ is perpendicular to both $\mathbf{r}$ and $\delta\boldsymbol{\phi}$, we can write $\delta\mathbf{r} = \delta\boldsymbol{\phi} \times \mathbf{r}$. Similarly, $\delta\mathbf{v} = \delta\boldsymbol{\phi} \times \mathbf{v}$. Therefore,

$$\sum_\alpha \left(\dot{\mathbf{p}}_\alpha \cdot \delta\boldsymbol{\phi} \times \mathbf{r}_\alpha + \mathbf{p}_\alpha \cdot \delta\boldsymbol{\phi} \times \mathbf{v}_\alpha \right) = 0.$$

Interchanging the dot and the cross we obtain

$$\sum_\alpha \delta\boldsymbol{\phi} \cdot \left(\mathbf{r}_\alpha \times \dot{\mathbf{p}}_\alpha + \dot{\mathbf{r}}_\alpha \times \mathbf{p}_\alpha \right) = \delta\boldsymbol{\phi} \cdot \sum_\alpha \frac{d}{dt}\left(\mathbf{r}_\alpha \times \mathbf{p}_\alpha \right) = 0.$$

Consequently, since $\delta\phi \neq 0$

$$\sum_\alpha \frac{d}{dt}(\mathbf{r}_\alpha \times \mathbf{p}_\alpha) = \frac{d}{dt}\sum_\alpha \mathbf{r}_\alpha \times \mathbf{p}_\alpha = \frac{d\mathbf{L}}{dt} = 0,$$

where $\mathbf{L}$ is the total angular momentum. That is, the isotropy of space leads to the conservation of angular momentum. (Be careful not to confuse the total vector angular momentum, $\mathbf{L}$, with the Lagrangian L.)

Exercise 1.22 Using the elementary mechanics approach, show that the angular momentum of a system of particles is constant if no net external torque acts on the system. Note that you must invoke Newton's third law in the strong form.

1.11.4 The conservation of energy and the work function

Consider the kinetic energy of a single particle. In introductory physics we learned the work–energy theorem, which states that the net work done on the particle by external forces will result in an equal increase in the kinetic energy of the particle. By definition, work is

$$W = \int \mathbf{F} \cdot \mathbf{ds}.$$

Consequently, the increase in kinetic energy of a particle acted upon by a force $\mathbf{F}$ as it goes from point 1 to point 2 is

$$T_2 - T_1 = \int_1^2 \mathbf{F} \cdot \mathbf{ds}.$$

If the force is *conservative*, the quantity $\mathbf{F} \cdot \mathbf{ds}$ is the differential of a scalar quantity denoted by U and called the "work function." The concept of work function has been replaced in modern terminology by potential energy (V) defined as the negative of the work function.

$$V = -U.$$

This means that if the force is *conservative* it can be expressed in terms of the gradient of a scalar function, thus

$$\mathbf{F} = -\nabla V.$$

Then the work energy theorem yields

$$T_2 - T_1 = \int_1^2 -\nabla V \cdot \mathbf{ds}.$$

But $\nabla V \cdot \mathbf{ds} = dV$, so

$$T_2 - T_1 = -(V_2 - V_1),$$

or

$$T_2 + V_2 = T_1 + V_1.$$

This equation expresses the conservation of mechanical energy; it indicates that there is a quantity that remains constant as the system is taken from one configuration to another under the action of conservative forces. This quantity is, of course, the total mechanical energy $E = T + V$.

The condition for a force to be conservative is that it is equal to the negative gradient of a scalar function. This is equivalent to requiring that the curl of the force be zero, i.e. $\nabla \times \mathbf{F} = 0$.

The preceding discussion of the conservation of energy was fairly elementary. We now consider energy from an advanced point of view and show that the conservation of energy follows from the homogeneity of time.

In general, the derivative of the Lagrangian $L = L(q, \dot{q}, t)$ with respect to time is

$$\frac{dL}{dt} = \sum_i \frac{\partial L}{\partial q_i} \frac{dq_i}{dt} + \sum_i \frac{\partial L}{\partial \dot{q}_i} \frac{d\dot{q}_i}{dt} + \frac{\partial L}{\partial t}.$$

By Lagrange's equation, $\frac{\partial L}{\partial q_i} = \frac{d}{dt} \frac{\partial L}{\partial \dot{q}_i}$, so

$$\frac{dL}{dt} = \sum_i \left[\dot{q}_i \frac{d}{dt} \left(\frac{\partial L}{\partial \dot{q}_i} \right) + \frac{\partial L}{\partial \dot{q}_i} \frac{d\dot{q}_i}{dt} \right] + \frac{\partial L}{\partial t}.$$

But note that

$$\frac{d}{dt} \sum_i \dot{q}_i \frac{\partial L}{\partial \dot{q}_i} = \sum_i \frac{d}{dt} \left(\dot{q}_i \frac{\partial L}{\partial \dot{q}_i} \right) = \sum_i \left[\dot{q}_i \frac{d}{dt} \left(\frac{\partial L}{\partial \dot{q}_i} \right) + \frac{d\dot{q}_i}{dt} \frac{\partial L}{\partial \dot{q}_i} \right],$$

so the summation in the preceding equation can be replaced by $\frac{d}{dt} \sum_i \dot{q}_i \frac{\partial L}{\partial \dot{q}_i}$, and

$$\frac{dL}{dt} = \frac{d}{dt} \sum_i \dot{q}_i \frac{\partial L}{\partial \dot{q}_i} + \frac{\partial L}{\partial t},$$

or

$$\frac{\partial L}{\partial t} = \frac{d}{dt} \left(L - \sum_i \dot{q}_i \frac{\partial L}{\partial \dot{q}_i} \right).$$

We now introduce a new quantity called the **energy function** h that is defined by

$$h = h(q, \dot{q}, t) = \sum_i \dot{q}_i \frac{\partial L}{\partial \dot{q}_i} - L,$$

so that

$$\frac{\partial L}{\partial t} = -\frac{dh}{dt}.$$

Note that h depends on the generalized positions and the generalized *velocities*.

Let us assume temporal homogeneity so that time does not appear explicitly in the Lagrangian. Then $\frac{\partial L}{\partial t} = 0$ and consequently

$$\frac{d}{dt}\left(L - \sum_i \dot{q}_i \frac{\partial L}{\partial \dot{q}_i} \right) = -\frac{dh}{dt} = 0.$$

That is, the energy function is constant. But does this imply that the *energy* $(E = T + V)$ is constant? To answer this question we note that the kinetic energy has the general form

$$T = T_0 + T_1 + T_2,$$

where T_0 does not depend on velocities, T_1 is a homogeneous function of the first degree in velocities and T_2 is a homogeneous function of second degree in velocities.[10] (In Cartesian coordinates, T_0 and T_1 are both zero, and T_2 is not only of second degree in velocities, but is actually a quadratic function of velocities, i.e. it contains $\dot{x}^2$ but not $\dot{x}\dot{y}$.)

Recall that a function $f(x_1, \ldots, x_n)$ is homogeneous of degree k if

$$f(\alpha x_1, \alpha x_2, \ldots, \alpha x_n) = \alpha^k f(x_1, \ldots, x_n).$$

Euler's theorem on homogeneous functions states that, if $f(x_i)$ is homogenous of degree k, then

$$\sum_i x_i \frac{\partial f}{\partial x_i} = kf.$$

Since the kinetic energy has the general form $T_0 + T_1 + T_2$, the Lagrangian also has this form: $L = L_0 + L_1 + L_2$. Then by Euler's theorem,

---

[10] This can be proved by starting with the kinetic energy in Cartesian coordinates and then using the transformation equations $x_i = x_i(q_1, q_2, \ldots, q_n, t)$, and the fact that $\dot{x}_i = \sum_j \frac{\partial x_i}{\partial q_j}\dot{q}_j + \frac{\partial x_i}{\partial t}$. See Problem 1.12 at the end of the chapter.

$$\sum_i \dot{q}_i \frac{\partial L_0}{\partial \dot{q}_i} = 0$$

$$\sum_i \dot{q}_i \frac{\partial L_1}{\partial \dot{q}_i} = L_1$$

$$\sum_i \dot{q}_i \frac{\partial L_2}{\partial \dot{q}_i} = 2L_2.$$

Therefore,

$$\sum_i \dot{q}_i \frac{\partial L}{\partial \dot{q}_i} = \sum_i \dot{q}_i \frac{\partial (L_0 + L_1 + L_2)}{\partial \dot{q}_i} = L_1 + 2L_2,$$

and

$$\sum_i \dot{q}_i \frac{\partial L}{\partial \dot{q}_i} - L = (L_1 + 2L_2) - (L_0 + L_1 + L_2) = L_2 - L_0.$$

If the transformation equations do not involve t explicitly, $T = T_2$, that is, T is a homogeneous function of second degree in the velocities. And if V does not depend on velocities, $L_0 = -V$. Then

$$h = T + V = E.$$

As long as these conditions are satisfied, the energy function is equal to the total energy and the total energy is conserved.

If we replace $\frac{\partial L}{\partial \dot{q}_i}$ by p_i, the energy function is transformed into the **Hamiltonian**, which is a function of the generalized coordinates and the generalized *momenta:*

$$H = H(q, p, t) = \sum_i \dot{q}_i p_i - L.$$

The energy function and the Hamiltonian are essentially the same thing, but they are expressed in terms of different variables, that is, $h = h(q, \dot{q}, t)$ and $H = H(q, p, t)$. (Whereas h depends on velocity, H depends on momentum.)

Exercise 1.23 (a) Show that $\mathbf{F} = -y\hat{i} + x\hat{j}$ is not conservative. (b) Show that $\mathbf{F} = y\hat{i} + x\hat{j}$ is conservative.

Exercise 1.24 Show that $\mathbf{F} = 3x^2 y\hat{i} + (x^3 + 1)\hat{j} + 9z^2\hat{k}$ is conservative.

Exercise 1.25 Prove that the curl of a conservative force is zero.

Exercise 1.26 Prove Euler's theorem. (Hint: Differentiate $f(tx, ty, tz) = t^k f(x, y, z)$ with respect to t, then let $t = 1$.)

Exercise 1.27 Show that $z^2 \ln(x/y)$ is homogeneous of degree 2.

Exercise 1.28 Show that $\sum_i \dot{q}_i \frac{\partial L_2}{\partial \dot{q}_i} = 2L_2$.

The virial theorem

The virial theorem states that for a bounded conservative system (such as a planet going around a star) the average value of the kinetic energy is proportional to the average value of the potential energy. (For the planet/star we find $\langle T \rangle = -(1/2) \langle V \rangle$.) We now prove the theorem.

In Cartesian coordinates the kinetic energy is a homogenous quadratic function depending only on the velocities, so Euler's theorem yields

$$\sum_\alpha \mathbf{v}_\alpha \cdot \frac{\partial T}{\partial \mathbf{v}_\alpha} = 2T.$$

As long as the potential energy does not depend on velocities, the momentum can be expressed as

$$\mathbf{p}_\alpha = \frac{\partial T}{\partial \mathbf{v}_\alpha},$$

and hence

$$2T = \sum_\alpha \mathbf{p}_\alpha \cdot \mathbf{v}_\alpha = \sum_\alpha \mathbf{p}_\alpha \cdot \frac{d\mathbf{r}_\alpha}{dt}.$$

But

$$\frac{d}{dt} \sum_\alpha \mathbf{p}_\alpha \cdot \mathbf{r}_\alpha = \sum_\alpha \mathbf{p}_\alpha \cdot \frac{d\mathbf{r}_\alpha}{dt} + \sum_\alpha \frac{d\mathbf{p}_\alpha}{dt} \cdot \mathbf{r}_\alpha.$$

Therefore

$$2T = \frac{d}{dt} \sum_\alpha \mathbf{p}_\alpha \cdot \mathbf{r}_\alpha - \sum_\alpha \frac{d\mathbf{p}_\alpha}{dt} \cdot \mathbf{r}_\alpha.$$

Now by Newton's second law, for a conservative system,

$$\frac{d\mathbf{p}_\alpha}{dt} = \mathbf{F}_\alpha = -\frac{\partial V}{\partial \mathbf{r}_\alpha},$$

so

$$2T = \frac{d}{dt} \sum_\alpha \mathbf{p}_\alpha \cdot \mathbf{r}_\alpha + \sum_\alpha \frac{\partial V}{\partial \mathbf{r}_\alpha} \cdot \mathbf{r}_\alpha.$$

Let us evaluate the average value of all the terms in this equation:

$$\langle 2T \rangle = \left\langle \frac{d}{dt} \sum_\alpha \mathbf{p}_\alpha \cdot \mathbf{r}_\alpha \right\rangle + \left\langle \sum_\alpha \mathbf{r}_\alpha \cdot \frac{\partial V}{\partial \mathbf{r}_\alpha} \right\rangle.$$

Recall that the time average of a function $f(t)$ is

$$\langle f(t) \rangle = \lim_{\tau \to \infty} \frac{1}{\tau} \int_0^\tau f(t)dt,$$

so

$$\left\langle \frac{d}{dt} \sum_\alpha \mathbf{p}_\alpha \cdot \mathbf{r}_\alpha \right\rangle = \lim_{\tau \to \infty} \frac{1}{\tau} \int_0^\tau \frac{d}{dt} \sum_\alpha \mathbf{p}_\alpha \cdot \mathbf{r}_\alpha dt = \lim_{\tau \to \infty} \frac{1}{\tau} \int_0^\tau d \left(\sum_\alpha \mathbf{p}_\alpha \cdot \mathbf{r}_\alpha \right)$$

$$= \lim_{\tau \to \infty} \frac{1}{\tau} \left[\sum_\alpha \mathbf{p}_\alpha \cdot \mathbf{r}_\alpha \right]_0^\tau$$

$$= \lim_{\tau \to \infty} \frac{\sum_\alpha \mathbf{p}_\alpha \cdot \mathbf{r}_\alpha \Big|_\tau - \sum_\alpha \mathbf{p}_\alpha \cdot \mathbf{r}_\alpha \Big|_0}{\tau} = 0,$$

where we assumed that $\sum_\alpha \mathbf{p}_\alpha \cdot \mathbf{r}_\alpha$ remains finite at all times. In other words, we are assuming the motion is bounded, so the numerator is finite but the denominator goes to infinity.

We are left with

$$2 \langle T \rangle = \left\langle \sum_\alpha \mathbf{r}_\alpha \cdot \frac{\partial V}{\partial \mathbf{r}_\alpha} \right\rangle.$$

If the potential energy is a homogeneous function of position of degree k, then by Euler's theorem the right-hand side is kV, and

$$2 \langle T \rangle = k \langle V \rangle.$$

Thus, for example, for a particle of mass m in a gravitational field, $V = -\frac{GMm}{r}$ so $k = -1$ and

$$2 \langle T \rangle = - \langle V \rangle.$$

Since $E = T + V$, we see that $\langle E \rangle = - \langle T \rangle$, indicating that the total energy is negative, and that the magnitude of the average kinetic energy is one half the magnitude of the average potential energy.

As another example, consider a mass on a spring. The potential energy is $V = (1/2)Kx^2$ so $k = 2$ and $2 \langle T \rangle = 2 \langle V \rangle$, that is, $\langle T \rangle = \langle V \rangle$.

1.12 Problems

1.1 There is a story about a physicist who got a traffic ticket for running a red light. Being a clever person the physicist proved to the judge that it was neither possible to stop nor to continue without either running the light or breaking the speed limit. In this problem we determine if the story is credible or is just a figment of the imagination of an overworked graduate

student. Assume the physicist is driving a car at a constant speed v_0, equal
to the speed limit. The car is a distance d from an intersection when the
light changes from green to yellow. The physicist has a reaction time τ and
the car brakes with a constant deceleration a_c. The light remains yellow for
a time t_y. Determine the conditions such that the car can neither stop nor
continue at v_0 without running the red light. Also, using realistic values
for τ, t_y, a_c and v_0 determine whether or not this situation could actually
arise. (Solve using simple kinematic relations.)

1.2 A cylinder of mass M and radius R is set *on end* on a table at a distance L
from the edge, as shown in Figure 1.11. A string is wound tightly around
the cylinder. The free end of the string passes over a frictionless pulley and
hangs off the edge of the table. A weight of mass m is attached to the free
end of the string. Determine the time required for the spool to reach the
edge of the table.

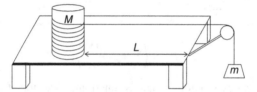

Figure 1.11   A cylinder on a frictionless table.

1.3 A string passes over a massless pulley. Each end is wound around a vertical
hoop, as shown in Figure 1.12. The hoops tend to descend, unwinding the
string, but if one hoop is much more massive than the other, it can cause
the lighter hoop to rise. The hoops have masses M_1 and M_2 and radii R_1
and R_2. Show that the tension in the string is $\tau = g M_1 M_2 (M_1 + M_2)^{-1}$.

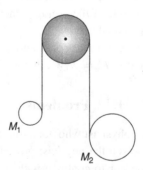

Figure 1.12   Two hoops hang from a pulley. The string unwinds as the hoops
descend.

1.4 The transformation equations for the bipolar coordinates η and ζ are

$$x = \frac{a \sinh \eta}{\cosh \eta - \cos \zeta},$$

$$y = \frac{a \sin \zeta}{\cosh \eta - \cos \zeta}.$$

(a) Show that the inverse transformation exists. (b) Obtain expressions for η and ζ in terms of x and y.

1.5 A disk of mass m and radius a rolls down a perfectly rough inclined plane of angle α. Determine the equations of motion and the constraints acting on the disk.

1.6 A pendulum of length l and mass m is mounted on a cart of mass M that is free to roll along a track. There is no friction. Write the Lagrangian for the pendulum/cart system.

1.7 Starting with the definition

$$T = \frac{1}{2} \sum_{i=1}^{N} m_i \left(\dot{x}_i^2 + \dot{y}_i^2 + \dot{z}_i^2 \right)$$

and the transformation equations, show that

$$T = \sum_{j=1}^{3N} \sum_{k=1}^{3N} \frac{1}{2} A_{jk} \dot{q}_j \dot{q}_k + \sum_{j=1}^{3N} B_j \dot{q}_j + T_0,$$

where

$$A_{jk} = \sum_{i=1}^{N} m_i \left(\frac{\partial x_i}{\partial q_j} \frac{\partial x_i}{\partial q_k} + \frac{\partial y_i}{\partial q_j} \frac{\partial y_i}{\partial q_k} + \frac{\partial z_i}{\partial q_j} \frac{\partial z_i}{\partial q_k} \right),$$

$$B_j = \sum_{i=1}^{N} m_i \left(\frac{\partial x_i}{\partial q_j} \frac{\partial x_i}{\partial t} + \frac{\partial y_i}{\partial q_j} \frac{\partial y_i}{\partial t} + \frac{\partial z_i}{\partial q_j} \frac{\partial z_i}{\partial t} \right),$$

$$T_0 = \sum_{i=1}^{N} \frac{1}{2} m_i \left[\left(\frac{\partial x_i}{\partial t} \right)^2 + \left(\frac{\partial y_i}{\partial t} \right)^2 + \left(\frac{\partial z_i}{\partial t} \right)^2 \right].$$

2

The calculus of variations

2.1 Introduction

The calculus of variations is a branch of mathematics which considers *extremal* problems; it yields techniques for determining when a particular definite integral will be a maximum or a minimum (or, more generally, the conditions for the integral to be "stationary"). The calculus of variations answers questions such as the following.

- What is the path that gives the shortest distance between two points in a plane? (A straight line.)
- What is the path that gives the shortest distance between two points on a sphere? (A geodesic or "great circle.")
- What is the shape of the curve of given length that encloses the greatest area? (A circle.)
- What is the shape of the region of space that encloses the greatest volume for a given surface area? (A sphere.)

The technique of the calculus of variations is to formulate the problem in terms of a definite integral, then to determine the conditions under which the integral will be maximized (or minimized). For example, consider two points (P_1 and P_2) in the x–y plane. These can be connected by an infinite number of paths, each described by a function of the form $y = y(x)$. Suppose we wanted to determine the equation $y = y(x)$ for the curve giving the shortest path between P_1 and P_2. To do so we note that the distance between the two end points is given by the definite integral

$$I = \int_{P_1}^{P_2} ds,$$

where ds is an infinitesimal distance along the curve. Our task is to determine the function $y = y(x)$ such that I is minimized.

As an analogy, let us recollect the introductory calculus method of finding the maximum or minimum (that is, a "stationary point") of some function. For example, consider an ordinary function, say $h(x, y)$, which we will assume represents a topological map with h being the altitude. Imagine a mountain whose height is greater than the height of any other point on the map. That is, at some point (say x_0, y_0) the height h is an absolute maximum. How can we determine x_0 and y_0? As you know from elementary calculus, the technique is to find the point at which the partial derivatives of h with respect to x and y are zero. (Clearly, as we go through a maximum, the slope of a curve changes from positive to negative, so it must be zero at the maximum itself.) But if the derivative of h with respect to x is zero, this means that points that are infinitesimally distant from the peak along the x direction will have the same value of h as the peak itself! That is, if $\partial h/\partial x = 0$, then there is no change in h for a small change in x. To resolve this conundrum, we note that near the peak, the function $h(x, y)$ can be expanded in a Taylor's series, thus,

$$h(x + dx, y + dy) = h(x_0, y_0) + dx \left.\frac{\partial h}{\partial x}\right|_{x_0, y_0} + dy \left.\frac{\partial h}{\partial y}\right|_{x_0, y_0}$$

$$+ \text{ second-order terms.}$$

The derivatives $\partial h/\partial x$ and $\partial h/\partial x$ on the right-hand side are zero so in the infinitesimal neighborhood of a stationary point the value of the function is, to first order, equal to its value at the stationary point. To find a change in h we need to go to the second-order terms. These terms will also tell us the nature of the stationary point: if they are positive the stationary point is a minimum, if they are negative it is a maximum and if they are positive in one direction and negative in another, the stationary point is a saddle point.

Of course, we are not dealing with the problem of finding a stationary point for a *function*, but rather for an *integral*. Furthermore, we are not varying the *coordinates*, but rather the *path* along which the integral is evaluated. Nevertheless, to find the stationary point for our integral we will use the concept that *to first order* the integral has the same value at all points in an infinitesimal neighborhood of the stationary point.

2.2 Derivation of the Euler–Lagrange equation

We now derive the basic equation of the calculus of variations. It is called the Euler–Lagrange equation, or simply the Euler equation. As we derive the

relation, we shall try to make things more concrete by considering a simple
example, namely the problem of finding the length of the shortest curve
between two points in a plane. The curve is described by the function $y = y(x)$. An infinitesimal element of the curve has length ds, where

$$ds = \sqrt{dx^2 + dy^2}.$$

The length of the curve is, then,

$$I = \int_i^f ds = \int_i^f \sqrt{dx^2 + dy^2} = \int_i^f \sqrt{1 + \left(\frac{dy}{dx}\right)^2}\, dx$$
$$= \int_i^f \sqrt{1 + y'^2}\, dx,$$

where the integral is taken between the end points (initial and final) of the
curve and the symbol y' is defined to be $y' = \frac{dy}{dx}$. (For convenience we denote
differentiation with respect to x by a prime.) See Figure 2.1.

Now suppose you want to find the curve that corresponds to the shortest path
length between the two end points. That means you want to find a function
$y = y(x)$ such that the integral I is a minimum.

The integral can be written in the following way

$$I = \int_i^f \Phi(y)\, dx, \qquad\qquad (2.1)$$

where Φ is a function of y. But y itself is a function (it is a function of x). So
Φ is a function of a function. We call it a *functional*.

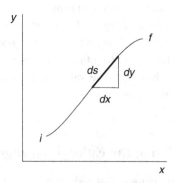

Figure 2.1   A curve $y = y(x)$ between points i and f.

The problem is to find the *function* $y(x)$ that minimizes the integral of the *functional* $\Phi(y)$. For the situation we are discussing, the functional is obviously

$$\Phi(y) = \sqrt{1 + y'^2}.$$

For different problems, as you can imagine, there are different functionals.

Here is a different problem. Determine the shape $z = z(x)$ of a frictionless wire such that a bead subjected to the gravitational force will slide from an initial point (x_i, z_i) to a final point (x_f, z_f) in the minimum amount of time. (This is a famous problem that was solved by Isaac Newton in a few hours; it is called the *Brachistochrone* problem.) In this problem we want to minimize the time. The time it takes an object to go from one point to another is given by the distance divided by the velocity. For a particle sliding down a wire the velocity can be determined from conservation of energy in the form $\frac{1}{2}mv^2 = mgh$, yielding

$$v = \sqrt{2gh},$$

where h is the distance below the initial point. If we measure z downwards from z_i we can write $v = \sqrt{2gz}$. The time to slide a distance ds along the curve is, then,

$$dt = \frac{ds}{\sqrt{2gz}} = \sqrt{\frac{dx^2 + dz^2}{2gz}} = \sqrt{\frac{1 + z'^2}{2gz}}\,dx.$$

The quantity to be minimized in this problem is

$$I = \int_i^f \sqrt{\frac{1 + z'^2}{2gz}}\,dx,$$

and the functional is

$$\Phi(z, z') = \sqrt{\frac{1 + z'^2}{2gz}}. \tag{2.2}$$

Example 2.1 *A geodesic is the shortest distance between two points on a spherical surface. This is a segment of a great circle. Eventually we will want to determine the equation for a great circle, but here we will simply determine the appropriate functional.*

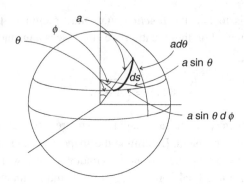

Figure 2.2   An element of path, ds, on the surface of a sphere.

Solution 2.1 *An element of path on a sphere of radius a is denoted ds in Figure 2.2. From the geometry of the sphere it is clear that*

$$ds^2 = a^2 d\theta^2 + a^2 \sin^2 \theta d\phi^2.$$

The length of a curve on the sphere is, therefore

$$I = \int ds = \int a d\theta \sqrt{1 + \sin^2 \theta \phi'^2},$$

where $\phi' = d\phi/d\theta$. The functional is

$$\Phi = \left(1 + \sin^2 \theta \phi'^2\right)^{1/2}.$$

The functionals obtained in these examples are representative of those found in calculus of variations problems. In general, the functionals we will be using depend on two functions, such as y and y', as well as the parameter x, thus:

$$\Phi = \Phi(x, y, y'). \tag{2.3}$$

Note that $y = y(x)$ and $y' = y'(x) = \frac{dy(x)}{dx}$. That is, x is the independent parameter and y and y' are functions of x. (Later we shall formulate problems in which the independent parameter is the time, t.)

What we have done so far is to describe how the functional is obtained for problems such as the shortest distance, the Brachistochrone and the geodesic. We have not, however, shown how to determine the function that minimizes the integral.

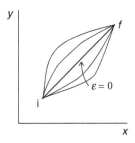

Figure 2.3   A family of curves $Y = y(x) + \epsilon\eta(x)$. The shortest distance curve is indicated by $\epsilon = 0$. Note that all the curves meet at the end points, so $\eta(x_i) = \eta(x_f) = 0$.

Let us now derive the condition for the integral to be stationary. I will use the shortest path between two points as an example, but the derivation is general. Imagine drawing a straight line between the points. As you know, this is the path that gives the shortest distance. Imagine drawing nearby paths between the same two points (i.e. paths that differ infinitesimally from the shortest path). See Figure 2.3. The nearby paths will differ slightly from the "true" or minimal path. If $y = y(x)$ is the equation of the minimal path, the equation of one of the nearby paths can be expressed as

$$Y(x, \epsilon) = y(x) + \epsilon\eta(x), \qquad (2.4)$$

where ϵ is a small quantity and $\eta(x)$ is an arbitrary function of x but which has an important condition on it, namely that $\eta(x)$ must be zero at the end points because at those points the paths all meet. For a given function $\eta(x)$, different values of ϵ will yield different paths, all belonging to same family of curves. (Selecting a different function $\eta(x)$ will generate another family of curves, each characterized by the value of ϵ.) Assuming a given $\eta(x)$, the path $Y(x)$ is a function of ϵ as well as x. That is why we wrote $Y = Y(x, \epsilon)$. The length of any one of these curves will, of course, depend on the value of ϵ and can be expressed as

$$I(\epsilon) = \int_{x_i}^{x_f} \Phi(x, Y, Y')dx,$$

where Φ has the form

$$\Phi = \sqrt{1 + Y'^2}.$$

We wrote I as a function only of ϵ because the dependence on x has been integrated out.

We want to determine the function $y(x)$ that makes the integral I stationary. In ordinary calculus we would set the differential to zero ($dI = 0$). But now we are asking which of the *functions* $y(x)$ makes I stationary, so we set the *variation* to zero ($\delta I = 0$). This is often called the "first variation" because δI is defined in terms of the Maclaurin expansion

$$I(\epsilon) = I(0) + \left[\frac{dI}{d\epsilon}\right]_{\epsilon=0} \epsilon + O(\epsilon^2).$$

Keeping only the first term in ϵ we define

$$\delta I(\epsilon) = I(\epsilon) - I(0) = \left[\frac{dI}{d\epsilon}\right]_{\epsilon=0} \epsilon.$$

Consequently, $\delta I = 0$ is equivalent to $\left[\frac{dI}{d\epsilon}\right]_{\epsilon=0} = 0$.
Inserting the integral expression for I, gives us

$$\left[\frac{d}{d\epsilon}\int_{x_i}^{x_f} \Phi(x, Y, Y')dx\right]_{\epsilon=0} = 0. \tag{2.5}$$

Y and Y' are both functions of ϵ. Taking the derivative under the integral we have

$$\int_{x_i}^{x_f}\left[\frac{\partial \Phi}{\partial Y}\frac{\partial Y}{\partial \epsilon} + \frac{\partial \Phi}{\partial Y'}\frac{\partial Y'}{\partial \epsilon}\right]_{\epsilon=0} dx = 0. \tag{2.6}$$

The second term can be integrated by parts as follows:

$$\int_{x_i}^{x_f}\left[\frac{\partial \Phi}{\partial Y'}\frac{\partial Y'}{\partial \epsilon}\right] dx = \int_{x_i}^{x_f}\frac{\partial \Phi}{\partial Y'}\frac{\partial}{\partial \epsilon}\left(\frac{dY}{dx}\right)dx = \int_{x_i}^{x_f}\frac{\partial \Phi}{\partial Y'}\frac{d}{dx}\left(\frac{\partial Y}{\partial \epsilon}\right)dx$$

$$= \int_{x_i}^{x_f}\frac{\partial \Phi}{\partial Y'}d\left(\frac{\partial Y}{\partial \epsilon}\right) = \frac{\partial \Phi}{\partial Y'}\frac{\partial Y}{\partial \epsilon}\Big|_{x_i}^{x_f} - \int_{x_i}^{x_f}\frac{d}{dx}\left(\frac{\partial \Phi}{\partial Y'}\right)\frac{\partial Y}{\partial \epsilon}dx.$$

But $Y = y(x) + \epsilon\eta(x)$ so $\frac{\partial Y}{\partial \epsilon} = \eta(x)$ and at the end points $\eta(x) = 0$ (because the curves meet there). Hence

$$\frac{\partial \Phi}{\partial Y'}\frac{\partial Y}{\partial \epsilon}\Big|_{x_i}^{x_f} = \frac{\partial \Phi}{\partial Y'}\left[\eta(x_f) - \eta(x_i)\right] = 0.$$

Consequently, Equation (2.6) is

$$\int_{x_i}^{x_f}\left[\left(\frac{\partial \Phi}{\partial Y} - \frac{d}{dx}\left[\frac{\partial \Phi}{\partial Y'}\right]\right)\left(\frac{\partial Y}{\partial \epsilon}\right)\right]_{\epsilon=0} dx = 0. \tag{2.7}$$

The integrand is the product of two functions. Since $\frac{\partial Y}{\partial \epsilon} = \eta(x)$, Equation (2.7) has the form

$$\int_a^b f(x)\eta(x)dx = 0,$$ (2.8)

where

$$f(x) = \frac{\partial \Phi}{\partial Y} - \frac{d}{dx}\left(\frac{\partial \Phi}{\partial Y'}\right).$$

The function $\eta(x)$ is arbitrary (except at the end points) so Equation (2.8) can be true if and only if $f(x) = 0$ for all values of x. You might think that requiring $f(x) = 0$ for all x is too restrictive, but it is easy to show that it is not. You can appreciate this by assuming $\eta(x)$ is zero everywhere except at some point between x_i and x_f, say at $x = \xi$. Next consider the integral

$$\int_{x_i=\xi-\varepsilon}^{x_f=\xi+\varepsilon} f(x)\eta(x)dx = 0,$$

where ε is a very small quantity. Since the integral is zero everywhere, it is zero in the range of integration. Furthermore, $f(x)$ is essentially constant over the small region of integration. Denote it by $f(\xi)$ and take it outside the integral sign,

$$f(\xi)\int_{x_i=\xi-\varepsilon}^{x_f=\xi+\varepsilon} \eta(x)dx = 0.$$

The integral is not zero, so $f(\xi)$ must be zero. But the point $x = \xi$ could be anywhere in the interval x_i to x_f, so $f(x)$ is zero at all points along the path.

Another argument that $\int_a^b f(x)\eta(x)dx = 0$ implies $f(x) = 0$ is to assume that $f(x)$ is not zero and use the fact that $\eta(x)$ is arbitrary. For example, we could require that $\eta(x)$ be negative everywhere that $f(x)$ is negative and that $\eta(x)$ be positive wherever $f(x)$ is positive. But then the product $f(x)\eta(x)$ would be positive everywhere and the condition (2.8) would not be met. Once again, we conclude that Equation (2.8) is true if and only if $f(x) = 0$. Consequently, Equation (2.7) reduces to

$$\left[\frac{\partial \Phi}{\partial Y} - \frac{d}{dx}\left(\frac{\partial \Phi}{\partial Y'}\right)\right]_{\epsilon=0} = 0,$$

which is equivalent to requiring that

$$\frac{\partial \Phi}{\partial y} - \frac{d}{dx}\left(\frac{\partial \Phi}{\partial y'}\right) = 0.$$ (2.9)

This is called the *Euler–Lagrange equation*.[1] It gives a condition that must be met if $y = y(x)$ is the path that minimizes the integral

$$\int_{x_i}^{x_f} \Phi(x, y, y') dx.$$

For example, in determining the equation for the minimal distance between two points in a plane, the functional was

$$\Phi(x, y, y') = \sqrt{1 + y'^2}.$$

Plugging this into the Euler–Lagrange equation we obtain

$$-\frac{d}{dx}\frac{\partial}{\partial y'}\left(1 + y'^2\right)^{\frac{1}{2}} = 0,$$

and this leads, after a bit of algebra, to

$$y = mx + b,$$

where m and b are constants. But $y = mx + b$ is the equation of a straight line!

Exercise 2.1 Show that the functional $\Phi = \sqrt{1 + y'^2}$ leads to $y = mx+b$.

Exercise 2.2 Find the equation for the shortest distance between two points in a plane using polar coordinates. Answer: $r\cos(\theta + \alpha) = C$, where α and C are constants.

Exercise 2.3 Determine and identify the curve $y = y(x)$ such that $\int_{x_1}^{x_2}[x(1 + y'^2)]^{1/2}dx$ is stationary. Answer: A parabola.

Exercise 2.4 Determine and identify the curve $y = y(x)$ such that $\int_{x_1}^{x_2}\frac{ds}{x}$ is stationary. Answer: A circle.

2.2.1 The difference between δ and d

The difference between δ and d is more than notational.

When applied to a *variable*, δ represents a *virtual displacement*, which we usually think of as a change in a coordinate carried out with time frozen. More generally, it is an infinitesimal change in some variable that is *virtual*; that

---

[1] You have met a variant of the Euler–Lagrange equation in Chapter 1 (Equation (1.16)).

is, carried out in an arbitrary but kinematically allowable manner while keeping the independent parameter constant. In contrast, we think of d as representing an actual change in a variable which occurs during some finite period of time.

When applied to a *function*, the symbol δ, which was introduced by Lagrange, represents the *variation* of the function, whereas d is the *differential* of the function. As a simple example, consider the function $F(x, \dot{x}, t)$, that is a function that depends on position, velocity and time. But position and velocity both depend on time. That is, time is the independent parameter, and we can write

$$F = F(x(t), \dot{x}(t), t).$$

According to the rules of calculus, the differential of F is

$$dF = \frac{\partial F}{\partial x}dx + \frac{\partial F}{\partial \dot{x}}d\dot{x} + \frac{\partial F}{\partial t}dt.$$

On the other hand, the variation of F is

$$\delta F = \frac{\partial F}{\partial x}\delta x + \frac{\partial F}{\partial \dot{x}}\delta\dot{x}.$$

The most significant difference between the differential and the variation is that the differential takes us to a different point *of the same function* whereas the variation leads to a *different function*. This is shown schematically in Figure 2.4, where we consider two functions of the independent variable x. These are $y = f(x)$, the "true" function (i.e. the function that minimizes the definite

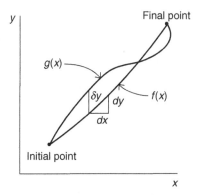

Figure 2.4   An illustration of the difference between δ and d.

integral) and $y = g(x) = f(x) + \epsilon \eta(x)$, which is a nearby function with the same end points, i.e. the "varied" function. The *variation* δy is given by

$$\delta y = g(x) - f(x).$$

That is, δy gives the difference between the true function and the varied function *at the same value of* x. On the other hand, if $y = f(x)$, then dy is the change in $f(x)$ due to a change in x. That is

$$dy = f(x + dx) - f(x).$$

The difference between δ and d becomes particularly important when applied to definite integrals because both kinds of changes are often acting simultaneously.

In the derivation of the Euler–Lagrange equation we used ϵ as a parameter that indicated whether or not we were on the "true" path (for which $\epsilon = 0$) or on one of the varied paths ($\epsilon \neq 0$). It is clear that the integral I giving the length of a path depends on ϵ. That is why we wrote $I = I(\epsilon)$. Our next step was to take the derivative of I with respect to ϵ and set it equal to zero. (The different values of ϵ take us to different curves, i.e. to different functions $y = y(x)$.)

We could also have expressed our problem in terms of variations, such as

$$\delta y = \eta(x)d\epsilon,$$

and

$$\delta y' = \eta'(x)d\epsilon.$$

These relations agree with the expression $Y(x) = y(x) + \epsilon \eta(x)$. If we recall that $Y(x)$ represents a family of curves $y(x, \epsilon)$, then replacing $Y(x)$ with $y(x, \epsilon)$ we have

$$\delta y = \delta[y(x, \epsilon)] = \left[\frac{dy(x, \epsilon)}{d\epsilon} \right]_{\epsilon=0} \delta\epsilon = \left[\frac{d}{d\epsilon} (y(x) + \epsilon \eta(x)) \right]_{\epsilon=0} \delta\epsilon = \eta(x)\delta\epsilon.$$

Since $I = I(y, y', x)$ we have

$$\delta I = \frac{\partial I}{\partial y}\delta y + \frac{\partial I}{\partial y'}\delta y'.$$

(Note that x is frozen.) But

$$\delta I = \delta \int_{x_1}^{x_2} \Phi(y, y', x)dx$$

$$= \int_{x_1}^{x_2} \delta\Phi dx$$

$$= \int_{x_1}^{x_2} \left(\frac{\partial \Phi}{\partial y} \delta y + \frac{\partial \Phi}{\partial y'} \delta y' \right) dx$$

$$= \int_{x_1}^{x_2} \left[\frac{\partial \Phi}{\partial y} \eta(x) d\epsilon + \frac{\partial \Phi}{\partial y'} \eta'(x) d\epsilon \right] dx.$$

Thus we could express the condition on I in either one of two ways, namely by setting

$$\left[\frac{dI}{d\epsilon} \right]_{\epsilon=0} = 0$$

as we did in Equation (2.7), or by stating that the variation δI is zero. These two statements are equivalent.

A final comment on notation. The "functional derivative" of $\Phi = \Phi(y, y', x)$ is defined by

$$\frac{\delta \Phi}{\delta y} = \frac{\partial \Phi}{\partial y} - \frac{d}{dx} \frac{\partial \Phi}{\partial y'}. \tag{2.10}$$

This is sometimes called the "variational derivative."

Exercise 2.5 Show that δ and d commute.

Exercise 2.6 Given

$$\delta \Phi = \frac{\partial \Phi}{\partial y} \delta y + \frac{\partial \Phi}{\partial y'} \delta y',$$

demonstrate that $\delta I = 0$ means the same thing as $\left[\frac{dI}{d\epsilon} \right]_{\epsilon=0} = 0$.

2.2.2 Alternate forms of the Euler–Lagrange equation

We derived the Euler–Lagrange equation in the form

$$\frac{\partial \Phi}{\partial y} - \frac{d}{dx} \left(\frac{\partial \Phi}{\partial y'} \right) = 0.$$

There are two variants of this expression. One of them is useful for problems in which Φ does not depend on y and the other is useful for problems in which Φ does not depend on x.

If Φ does not depend on y, the Euler–Lagrange equation reduces to

$$\frac{d}{dx} \left(\frac{\partial \Phi}{\partial y'} \right) = 0,$$

from which we conclude that

$$\frac{\partial \Phi}{\partial y'} = \text{constant.}$$

(This was the situation when we solved the "shortest distance in a plane" problem.)

Next assume that Φ does not depend on x. Then, since $\Phi = \Phi(y, y')$, the partial derivative of Φ with respect to x is zero and the total derivative of Φ with respect to x will be

$$\frac{d\Phi}{dx} = \frac{\partial \Phi}{\partial y}\frac{dy}{dx} + \frac{\partial \Phi}{\partial y'}\frac{dy'}{dx}.$$

But $\frac{dy'}{dx} = \frac{d}{dx}\frac{dy}{dx} = y''$, so

$$\frac{d\Phi}{dx} = \frac{\partial \Phi}{\partial y}\frac{dy}{dx} + y''\frac{\partial \Phi}{\partial y'} = y'\frac{\partial \Phi}{\partial y} + y''\frac{\partial \Phi}{\partial y'}. \tag{2.11}$$

If we multiply the Euler–Lagrange equation by y' we obtain

$$y'\frac{\partial \Phi}{\partial y} - y'\frac{d}{dx}\left(\frac{\partial \Phi}{\partial y'}\right) = 0.$$

Add $y''\frac{\partial \Phi}{\partial y'}$ to both sides:

$$y'\frac{\partial \Phi}{\partial y} - y'\frac{d}{dx}\left(\frac{\partial \Phi}{\partial y'}\right) + y''\frac{\partial \Phi}{\partial y'} = y''\frac{\partial \Phi}{\partial y'}$$

$$y'\frac{\partial \Phi}{\partial y} + y''\frac{\partial \Phi}{\partial y'} = y''\frac{\partial \Phi}{\partial y'} + y'\frac{d}{dx}\left(\frac{\partial \Phi}{\partial y'}\right).$$

The left-hand side is just $\frac{d\Phi}{dx}$ (see Equation (2.11)) and the right-hand side is $\frac{d}{dx}\left(y'\frac{\partial \Phi}{\partial y'}\right)$, so

$$\frac{d\Phi}{dx} = \frac{d}{dx}\left(y'\frac{\partial \Phi}{\partial y'}\right),$$

or

$$\frac{d}{dx}\left(\Phi - y'\frac{\partial \Phi}{\partial y'}\right) = 0.$$

Therefore,

$$\Phi - y'\frac{\partial \Phi}{\partial y'} = \text{constant.} \tag{2.12}$$

Example 2.2 *The Brachistochrone problem: The problem is to determine the shape of a wire such that a bead on the wire will slide from some initial point to some lower final point in the least amount of time. Note that the two end points (x_1, z_1) and (x_2, z_2) are fixed. In Section 2.2 we obtained the functional. We now solve for the curve $z = z(x)$.*

Solution 2.2 *The functional was given by Equation (2.2) as*

$$\Phi(z, z') = \sqrt{\frac{1 + z'^2}{2gz}}.$$

Note that Φ does not depend on x. Therefore,

$$\Phi - z' \frac{\partial \Phi}{\partial z'} = constant.$$

The factor $2g$ can be discarded. Carrying out the mathematics, we obtain

$$\sqrt{\frac{1 + z'^2}{z}} - \frac{z'^2}{\sqrt{z}\sqrt{1 + z'^2}} = constant = 1/\sqrt{C}. \tag{2.13}$$

This can be expressed as

$$z(1 + z'^2) = C. \tag{2.14}$$

The details are left as an exercise. We can obtain a solution to this differential equation by setting

$$z' = \cot \theta.$$

Then $1 + z'^2 = 1/\sin^2 \theta$, and our differential equation can be written

$$z(1 + \cot^2 \theta) = z/\sin^2 \theta = C,$$

or

$$z = C \sin^2 \theta.$$

Next note that

$$\frac{dx}{d\theta} = \frac{dx}{dz}\frac{dz}{d\theta} = \frac{1}{z'}\frac{dz}{d\theta} = \tan \theta \frac{dz}{d\theta} = \tan \theta \frac{d}{d\theta}(C2 \sin^2 \theta) = C(1 - \cos 2\theta).$$

Therefore

$$x = \int C(1 - \cos 2\theta) d\theta = C\left[\theta - \frac{\sin 2\theta}{2}\right].$$

Letting $C = 2A$ and $2\theta = \phi$ we can write the following parametric equations for x and z:

$$x = A(\phi - \sin \phi)$$
$$z = A(1 - \cos \phi).$$

These are the parametric equations for a cycloid. Consequently, the bead slides in minimum time from (x_1, z_1) to (x_2, z_2) if the wire is bent into a cycloid.

Exercise 2.7 Show the equivalence of the following two forms of the Euler–Lagrange equation

$$\frac{\partial \Phi}{\partial y} - \frac{d}{dx}\frac{\partial \Phi}{\partial y'} = 0$$

and

$$\frac{\partial \Phi}{\partial x} - \frac{d}{dx}\left(\Phi - y'\frac{\partial \Phi}{\partial y'}\right) = 0,$$

assuming $y' \neq 0$.

Exercise 2.8 Obtain Equation (2.14) starting with Equation (2.13).

Exercise 2.9 Determine the minimum time for a bead to slide down a wire from the point $(-\pi, 2)$ to $(0, 0)$ (meters). Answer: $\pi/\sqrt{g}$.

2.3 Generalization to several dependent variables

We have been considering problems whose functionals have the form $\Phi = \Phi(x, y, y')$. In this section we generalize to functionals that depend on n dependent variables, $y_1, y_2, \ldots, y_n$ and their derivatives, $y'_1, y'_2, \ldots, y'_n$. We assume the variables are independent.[2]

The functional for the system is now $\Phi(y_1, y_2, \ldots, y_n, y'_1, y'_2, \ldots, y'_n, x)$. We will show that this leads to n Euler–Lagrange equations of the form:

$$\frac{d}{dx}\left(\frac{\partial \Phi}{\partial y'_i}\right) - \frac{\partial \Phi}{\partial y_i} = 0, \quad i = 1, \ldots, n.$$

---

[2] This section is included for the sake of completeness. The procedure is a generalization of the the simpler situation described in the the previous section (2.2). You may skip this section without loss of continuity, but note that some of the results obtained here are used in Section 3.4.

The derivation is similar to that presented in the preceding section, but is more general.

Recall that our intention is to determine the functions $y_1(x)$, $y_2(x)$, ... which will minimize (or maximize) the definite integral

$$I = \int_{x_i}^{x_f} \Phi(y_1, \ldots, y_n, y_1', \ldots, y_n', x) dx.$$

That is, we want to obtain the functions $y_i(x)$ such that

$$\delta I = 0.$$

As before, there are an infinite number of paths between the end points. The paths that are infinitesimally near the minimal or true path can be described by

$$Y_i(x) = y_i(x) + \epsilon \eta_i(x).$$

The integral I depends on which path is chosen, so it can be considered a function of ϵ.

The variation of I is

$$\delta I = \delta \int_{x_i}^{x_f} \Phi dx = \int_{x_i}^{x_f} \delta \Phi dx = \int_{x_i}^{x_f} \sum_i \left(\frac{\partial \Phi}{\partial Y_i} \delta Y_i + \frac{\partial \Phi}{\partial Y_i'} \delta Y_i' \right) dx.$$

Considering Y_i and Y_i' as functions of ϵ, we write

$$\delta I = \left(\frac{\partial I}{\partial \epsilon} \right)_{\epsilon=0} d\epsilon = \left[\int_{x_i}^{x_f} \sum_i \left(\frac{\partial \Phi}{\partial Y_i} \frac{\partial Y_i}{\partial \epsilon} + \frac{\partial \Phi}{\partial Y_i'} \frac{\partial Y_i'}{\partial \epsilon} \right) dx \right]_{\epsilon=0} d\epsilon.$$

The second term can be integrated by parts. (Note that the mathematical procedure here is nearly identical to that from Equation (2.6) to Equation (2.7).) We have

$$\int_{x_i}^{x_f} \frac{\partial \Phi}{\partial Y_i'} \frac{\partial Y_i'}{\partial \epsilon} d\epsilon dx = \int_{x_i}^{x_f} \frac{\partial \Phi}{\partial Y_i'} \frac{\partial^2 Y_i}{\partial \epsilon \partial x} dx d\epsilon = \int_{x_i}^{x_f} \frac{\partial \Phi}{\partial Y_i'} \frac{\partial}{\partial \epsilon} \left(\frac{\partial Y_i}{\partial x} \right) dx \, d\epsilon$$

$$= \frac{\partial \Phi}{\partial Y_i'} \frac{\partial Y_i}{\partial \epsilon} \bigg|_{x_1}^{x_2} - \int_{x_i}^{x_f} \frac{\partial Y_i}{\partial \epsilon} \frac{d}{dx} \frac{\partial \Phi}{\partial Y_i'} dx d\epsilon$$

$$= - \int_{x_i}^{x_f} \frac{\partial Y_i}{\partial \epsilon} \frac{d}{dx} \frac{\partial \Phi}{\partial Y_i'} dx d\epsilon,$$

where we have used the fact that $\frac{\partial Y_i}{\partial \epsilon} \big|_{x_1}^{x_2} = 0$ because all the curves pass through the end points.

Consequently,

$$
\delta I = \int_{x_i}^{x_f} \sum_i \left(\frac{\partial \Phi}{\partial Y_i} \frac{\partial Y_i}{\partial \epsilon} - \frac{\partial Y_i}{\partial \epsilon} \frac{d}{dx} \frac{\partial \Phi}{\partial Y_i'} \right) d\epsilon dx
$$

$$
= \int_{x_i}^{x_f} \sum_i \left(\frac{\partial \Phi}{\partial Y_i} - \frac{d}{dx} \frac{\partial \Phi}{\partial Y_i'} \right) \frac{\partial Y_i}{\partial \epsilon} d\epsilon dx
$$

$$
= \int_{x_i}^{x_f} \sum_i \left(\frac{\partial \Phi}{\partial Y_i} - \frac{d}{dx} \frac{\partial \Phi}{\partial Y_i'} \right) \delta Y_i dx.
$$

When $\epsilon = 0$, $\delta I = 0$ because I is minimized along the path $y_i = y_i(x)$. So

$$
[\delta I]_{\epsilon=0} = 0 = \int_{x_i}^{x_f} \sum_i \left(\frac{\partial \Phi}{\partial y_i} - \frac{d}{dx} \frac{\partial \Phi}{\partial y_i'} \right) \delta y_i dx. \qquad (2.15)
$$

Since the δy_i s are independent, the only way this expression can be zero is for all the terms in the parenthesis to be zero. Hence,

$$
\frac{\partial \Phi}{\partial y_i} - \frac{d}{dx} \left(\frac{\partial \Phi}{\partial y_i'} \right) = 0, \quad i = 1, \ldots, n, \qquad (2.16)
$$

as expected.

2.4 Constraints

Many problems in the calculus of variations involve constraints or, as they are sometimes called, "auxiliary conditions." For example, a particle might be constrained to the $z = 0$ plane or to move along a specific curve.

2.4.1 Holonomic constraints

Let us begin by considering holonomic constraints. Recall from Section 1.4 that a holonomic constraint is a relationship between the dependent variables that can be expressed as a function that is equal to zero. In the presence of a constraint, the dependent variables are not independent of one another, being related by the auxiliary conditions. For example, let us consider a problem described by $\Phi = \Phi(y_1, \ldots, y_n; y_1', \ldots, y_n'; x)$ where the n dependent variables are related by m constraints of the form

$$
f_1(y_1, y_2, \ldots, y_n, x) = 0,
$$

$$
\vdots
$$

$$
f_m(y_1, y_2, \ldots, y_n, x) = 0.
$$

Each constraint reduces by one the number of degrees of freedom. If there are n dependent coordinates and m constraints we can use the constraints to reduce the number of degrees of freedom to $n - m$ and then apply $n - m$ Euler–Lagrange equations to solve the problem. This is conceptually simple, but may be fairly complicated in practice. A better approach is to keep all n variables and use the method of Lagrange multipliers. We now consider this method (often referred to as Lagrange's λ-method). We begin by taking the variation of the constraint equations, thus:

$$\delta f_1 = \frac{\partial f_1}{\partial y_1} \delta y_1 + \frac{\partial f_1}{\partial y_2} \delta y_2 + \cdots + \frac{\partial f_1}{\partial y_n} \delta y_n = 0,$$

$$\vdots$$

$$\delta f_m = \frac{\partial f_m}{\partial y_1} \delta y_1 + \frac{\partial f_m}{\partial y_2} \delta y_2 + \cdots + \frac{\partial f_m}{\partial y_n} \delta y_n = 0. \tag{2.17}$$

Next, we multiply each of these relations by an "undetermined multiplier" λ_i and sum them to obtain

$$\lambda_1 \delta f_1 + \lambda_2 \delta f_2 + \cdots + \lambda_m \delta f_m = 0. \tag{2.18}$$

The sum is zero because each δf is zero.

We also know that the variation of the functional Φ vanishes at the stationary point. Assuming Φ is a function of coordinates only,

$$\delta \Phi = \frac{\partial \Phi}{\partial y_1} \delta y_1 + \frac{\partial \Phi}{\partial y_2} \delta y_2 + \cdots + \frac{\partial \Phi}{\partial y_n} \delta y_n = 0. \tag{2.19}$$

The sum of Equations (2.18) and (2.19) must also be zero, so we have

$$0 = \frac{\partial \Phi}{\partial y_1} \delta y_1 + \frac{\partial \Phi}{\partial y_2} \delta y_2 + \cdots + \frac{\partial \Phi}{\partial y_n} \delta y_n$$
$$+ \lambda_1 \left(\frac{\partial f_1}{\partial y_1} \delta y_1 + \frac{\partial f_1}{\partial y_2} \delta y_2 + \cdots + \frac{\partial f_1}{\partial y_n} \delta y_n \right)$$
$$+ \cdots + \lambda_m \left(\frac{\partial f_m}{\partial y_1} \delta y_1 + \frac{\partial f_m}{\partial y_2} \delta y_2 + \cdots + \frac{\partial f_m}{\partial y_n} \delta y_n \right). \tag{2.20}$$

The expression is complicated so, to illustrate the procedure, I will assume that there is a single constraint. (The generalization to m constraints is left as a problem.) If there is only one constraint, Equation (2.18) can be written as

$$\lambda \delta f = \lambda \left(\frac{\partial f}{\partial y_1} \delta y_1 + \frac{\partial f}{\partial y_2} \delta y_2 + \cdots + \frac{\partial f}{\partial y_n} \delta y_n \right) = 0, \tag{2.21}$$

and Equation (2.20) reduces to

$$\frac{\partial \Phi}{\partial y_1}\delta y_1 + \frac{\partial \Phi}{\partial y_2}\delta y_2 + \cdots + \frac{\partial \Phi}{\partial y_n}\delta y_n$$
$$+ \lambda \left(\frac{\partial f}{\partial y_1}\delta y_1 + \frac{\partial f}{\partial y_2}\delta y_2 + \cdots + \frac{\partial f}{\partial y_n}\delta y_n \right) = 0.$$

Re-arranging yields

$$\sum_{i=1}^{n} \left(\frac{\partial \Phi}{\partial y_i} + \lambda \frac{\partial f}{\partial y_i} \right) \delta y_i = 0.$$

Since we introduced λ arbitrarily, we are free to give it any value we wish. Let us select λ such that the nth term of the sum is zero. That is, the value of λ will be determined by requiring that

$$\frac{\partial \Phi}{\partial y_n} + \lambda \frac{\partial f}{\partial y_n} = 0. \tag{2.22}$$

Then the nth term of the summation is eliminated, leaving us with

$$\sum_{i=1}^{n-1} \left(\frac{\partial \Phi}{\partial y_i} + \lambda \frac{\partial f}{\partial y_i} \right) \delta y_i = 0.$$

Now all the δys are independent and the summation is zero if and only if each coefficient of a variation δy_i $(i = 1, \ldots, n-1)$ is zero. That is,

$$\frac{\partial \Phi}{\partial y_i} + \lambda \frac{\partial f}{\partial y_i} = 0 \quad i = 1, 2, \ldots, n-1. \tag{2.23}$$

Note that we have not eliminated any variables. We can sum Equations (2.21) and (2.23) to obtain

$$\delta \Phi + \lambda \delta f = 0.$$

This can be written as

$$\delta(\Phi + \lambda f) = 0, \tag{2.24}$$

because $\delta(\lambda f) = f\delta\lambda + \lambda\delta f = \lambda\delta f$ since $f = 0$.

  Equation (2.24) states that the variation of $\Phi + \lambda f$ is zero for arbitrary variations in the generalized coordinates. Thus, we have reformulated the problem. Originally we had $\delta \Phi = 0$ subject to the constraint $f(y_1 \ldots y_n) = 0$, but now we have $\delta(\Phi + \lambda f) = 0$ with no constraints on the coordinates. The original problem had $n-1$ degrees of freedom whereas the reformulated problem treats all the coordinates as unconstrained and introduces another variable λ giving $n+1$ degrees of freedom. Note further that $\delta(\Phi + \lambda f) = 0$ yields n equations, and $f = 0$ is one more equation. Therefore we have $n+1$ equations in $n+1$ unknowns.

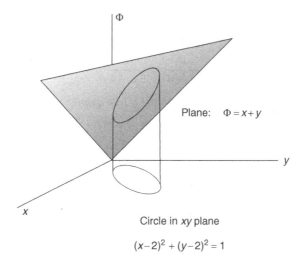

Plane: $\Phi = x + y$

Circle in xy plane

$(x-2)^2 + (y-2)^2 = 1$

Figure 2.5 The circle in the xy plane is projected upwards onto the shaded surface.

Example 2.3 *As a very simple application of the foregoing, consider the plane* $\Phi(x, y) = x + y$. *A circle in the xy plane is projected onto the Φ plane, as shown in Figure 2.5. Determine the location of the maximum and minimum points of the projected circle, given the constraint* $(x - 2)^2 + (y - 2)^2 = 1$.

Solution 2.3 *Given*

$$\Phi(x, y) = x + y,$$
$$f(x, y) = (x - 2)^2 + (y - 2)^2 - 1 = 0,$$

the condition

$$\delta(\Phi + \lambda f) = 0,$$

when inserted into the Euler–Lagrange equation, leads to the following three equations

$$\frac{\partial}{\partial x}(\Phi + \lambda f) = 0 \Rightarrow 1 + \lambda(2(x - 2)) = 0,$$
$$\frac{\partial}{\partial y}(\Phi + \lambda f) = 0 \Rightarrow 1 + \lambda(2(y - 2)) = 0,$$
$$\frac{\partial}{\partial \lambda}(\Phi + \lambda f) = 0 \Rightarrow (x - 2)^2 + (y - 2)^2 - 1 = 0.$$

The third equation is simply the constraint. The first and second equations can be solved for λ*. Equating the two expressions for* λ *we obtain*

$$\frac{1}{2x-4} = \frac{1}{2y-4}$$

which yields $x = y$*. Inserting this into the third equation we obtain*

$$2(x-2)^2 = 1$$

from which

$$x = 2 \pm 1/\sqrt{2}$$

and consequently, the extrema are at

$$(1.29, 1.29) \textit{ and } (2.71, 2.71).$$

2.4.2　Non-holonomic constraints

Differential constraints

Suppose the constraint is not an algebraic relation between the variables, but a differential relation. For example, the "rolling without slipping" constraint is expressed as a relation between the linear displacement and the angular displacement in the form

$$ds = rd\theta.$$

Dividing by dt you note that this is a relationship between velocities rather than coordinates. Now obviously if you can integrate the relation between differentials, you will get a holonomic constraint, and you can proceed as before. If the constraint is truly non-holonomic, all is not lost because for *some* non-holonomic constraints one can apply the Lagrangian λ-method described above.

Assume the constraint is expressed in the form

$$A_1 \delta y_1 + A_2 \delta y_2 + \cdots + A_n \delta y_n = 0.$$

There are two things to be noted here. First of all, the partial derivatives that appeared in Equation (2.17) are replaced by the coefficients A_i which are functions of the dependent variables but they are not derivatives of some function. Secondly, we cannot eliminate a variable because we do not have the equations between them that we need to eliminate one variable in terms of the others. Nevertheless, the Lagrangian λ-method can be applied. Define the quantity δf:

$$\delta f = A_1 \delta y_1 + A_2 \delta y_2 + \cdots + A_n \delta y_n = 0.$$

Multiply δf by λ and add it to the variation of the functional Φ:

$$\delta\Phi + \lambda\delta f = 0.$$

Now, once again, you can treat all the y_is as independent.

If the auxiliary conditions are time dependent (rheonomous) the procedure is a bit more complicated and we will not consider it. The interested student may read Section 2.13 of Lanczos[3] for a brief discussion.

Isoperimetric constraints

Some constraints are expressed as integrals. Such a constraint is called an "isoperimetric condition" because the most famous problem of this type is the "Queen Dido" problem of finding the curve of given perimeter that encloses the greatest area.[4] In general the constraint is given as a definite integral which has a fixed given value, thus

$$\int_{x_1}^{x_2} f(x, y, y')dx = C, \tag{2.25}$$

where C is known. Such a problem is solvable using Lagrange's λ-method.

Assume the integral to be maximized is

$$I = \int_{x_1}^{x_2} \Phi(x, y, y')dx. \tag{2.26}$$

For simplicity let us consider a system subjected to a single constraint. (The generalization to several conditions of the form (2.25) is straightforward.) The variation of the condition (2.25) is

$$\delta \int_{x_1}^{x_2} f(x, y, y')dx = \int_{x_1}^{x_2} \left(\frac{\partial f}{\partial y}\delta y + \frac{\partial f}{\partial y'}\delta y'\right) dx = 0. \tag{2.27}$$

Multiply Equation (2.27) by the undetermined constant λ and add it to δI to obtain

$$\delta \int_{x_1}^{x_2} (\Phi + \lambda f) \, dx = 0.$$

Thus the Lagrangian λ-method implies that the integral

$$\int_{x_1}^{x_2} (\Phi + \lambda f) \, dx$$

[3] Lanczos, *op. cit.*

[4] The legend tells us that Queen Dido, escaping from her brother, went to Carthage. She offered to buy land, but the ruler of the town haughtily told her she could only have as much land as she could enclose in the hide of an ox. Dido cut the hide of the largest ox she could find into the thinnest strip possible. The problem is to show that, given the length of the strip, the largest area is enclosed by a circle.

is stationary and hence the integrand satisfies the Euler–Lagrange equation. Note that the constraint and the integral of the functional have the same limits of integration.

Example 2.4 *Determine the shape of the curve of length L that encloses the greatest area.*

Solution 2.4 *Recall from the calculus that the area enclosed by a curve C is given by*

$$Area = A = \frac{1}{2}\oint_C (x\,dy - y\,dx).$$

Expressing x and y parametrically as x(t) and y(t) we write

$$\Phi = \frac{1}{2}\oint_C \left(x\frac{dy}{dt} - y\frac{dx}{dt} \right) dt = \frac{1}{2}\oint_C (xy' - yx')\,dt.$$

(Here t is just a parameter – it is not the time.) The constraint that the curve has length L is

$$f = \oint_C \sqrt{x'^2 + y'^2}\,dt - L = 0.$$

Note that $\Phi = \Phi(x, y, x', y', t)$ *with t as the independent parameter. Since* $\Phi + \lambda f$ *is stationary, it obeys the following Euler–Lagrange equations:*

$$\frac{\partial(\Phi + \lambda f)}{\partial x} - \frac{d}{dt}\left[\frac{\partial(\Phi + \lambda f)}{\partial x'} \right] = 0,$$

$$\frac{\partial(\Phi + \lambda f)}{\partial y} - \frac{d}{dt}\left[\frac{\partial(\Phi + \lambda f)}{\partial y'} \right] = 0,$$

yielding

$$\frac{1}{2}y' - \frac{d}{dt}\left[-\frac{1}{2}y + \frac{\lambda x'}{\sqrt{x'^2 + y'^2}} \right] = 0,$$

$$-\frac{1}{2}x' - \frac{d}{dt}\left[\frac{1}{2}x + \frac{\lambda y'}{\sqrt{x'^2 + y'^2}} \right] = 0.$$

Integrating, we obtain

$$\frac{1}{2}y + \frac{1}{2}y - \frac{\lambda x'}{\sqrt{x'^2 + y'^2}} = C_2,$$

$$-\frac{1}{2}x - \frac{1}{2}x - \frac{\lambda y'}{\sqrt{x'^2 + y'^2}} = -C_1,$$

where C_1 and C_2 are constants of integration. Rearranging we write

$$y - C_2 = \frac{\lambda x'}{\sqrt{x'^2 + y'^2}},$$

$$x - C_1 = \frac{\lambda y'}{\sqrt{x'^2 + y'^2}}.$$

Consequently, squaring and adding yields

$$(x - C_1)^2 + (y - C_2)^2 = \frac{\lambda^2}{x'^2 + y'^2}(x'^2 + y'^2) = \lambda^2,$$

which is the equation of a circle of radius λ with center at (C_1, C_2).

2.5 Problems

2.1 Generalize the Lagrange λ-method to a system with n coordinates and m constraints $(m < n)$ and show that if $\Phi = \Phi(y_i, x)$

$$\delta(\Phi + \lambda_1 f_1 + \cdots + \lambda_m f_m) = 0,$$

where the λs are obtained from

$$\frac{\partial \Phi}{\partial y_i} + \lambda_1 \frac{\partial f_1}{\partial y_i} + \cdots + \lambda_m \frac{\partial f_m}{\partial y_i} = 0, \qquad i = n - m + 1, \ldots, n).$$

2.2 Determine the equation of the curve giving the shortest distance between two points on the surface of a cone. Let $r^2 = x^2 + y^2$ and $z = r \cot \alpha$.

2.3 Determine the equation for shortest curve between two points on the surface of cylinder $(r = 1 + \cos \theta)$.

2.4 Consider a solid of revolution of a given height. Determine the shape of the solid if it has the minimum moment of inertia about its axis.

2.5 Consider the variational problem for variable end points. (a) Let

$$I = \int_a^b \Phi(x, y, y') dx,$$

where $y(b)$ is arbitrary. Show that

$$\delta I = \eta \frac{\partial \Phi}{\partial y'}\Big|_a^b + \int_a^b \left(\frac{\partial \Phi}{\partial y} - \frac{d}{dx} \frac{\partial \Phi}{\partial y'} \right) \eta dx,$$

with the condition

$$\frac{\partial \Phi}{\partial y'}\Big|_{x=b} = 0.$$

Figure 2.6   A frictionless tunnel for rapid transits between two points on Earth.

(b) Assume y is fixed at $x = a$ but the other end point can lie anywhere on a curve $g(x, y) = 0$. Show that in addition to the Euler–Lagrange condition we have

$$\left(\Phi - y'\frac{\partial\Phi}{\partial y'}\right)\frac{\partial g}{\partial y} - \frac{\partial\Phi}{\partial y'}\frac{\partial g}{\partial x} = 0.$$

2.6 A frictionless tunnel is dug through the Earth from point a to point b. An object dropped in the opening will slide through the tunnel under the action of gravity and emerge at the other end with zero velocity. The gravitational potential at a point inside the Earth is $\phi = -Gm_s/r$ where m_s is the mass of the material enclosed by a sphere of radius r. (We are assuming a constant density Earth.) Use polar coordinates and find the equation of the path that minimizes the time $\int dt$. Determine the transit time.[5] See Figure 2.6.

2.7 Two rings of radius a are placed a distance $2b$ apart. (The centers of the rings lie along the same line and the planes of the rings are perpendicular to this line.) Find the shape of a soap film formed between the rings, noting that the film will have minimal surface area.

2.8 A particle of mass m is in a two-dimensional force field given by

$$\mathbf{F} = -G\frac{Mm}{r^2}\hat{\mathbf{r}}.$$

The curve requiring the minimum time for the particle to fall from one point to another is the solution of the differential equation

$$\frac{dr}{d\theta} = f(r).$$

Determine $f(r)$.

2.9 Consider the function

$$f(x, y, z) = x^2 + 2y^2 + 3z^2 + 2xy + 2xz,$$

[5] See P. W. Cooper, Through the Earth in forty minutes, *Am. J. Phys.*, **34**, 68 (1966) and G. Venezian, Terrestrial brachistochrone, *Am. J. Phys.*, **34**, 701 (1966).

subject to the condition

$$x^2 + y^2 + z^2 = 1.$$

What is the minimum value of $f(x, y, z)$?

2.10 The speed of light in a medium of index of refraction n is $v = c/n = ds/dt$. The time for light to travel from point A to point B is

$$\int_A^B \frac{ds}{v}.$$

Obtain the law of reflection and the law of refraction (Snell's law) by using Fermat's principle of least time. Using polar coordinates, show that if n is proportional to $1/r^2$, that the path followed by a ray of light is given by $\sin(\theta + c) = kr$, where c and k are constants.

2.11 (This is known as "Newton's problem.") Consider a solid of revolution generated by a curve from the origin to a point B in the first quadrant by rotating the curve about the x axis. It is desired to determine the shape of this curve if the air resistance on the solid is a minimum. (The solid is moving towards the left.) Isaac Newton assumed the resistance was given by the expression

$$2\pi \rho V^2 \int y \sin^2 \Psi \, dy,$$

where $\Psi = y' =$ the slope of the curve, ρ is the air density and V^2 is the velocity. (This is not a very good expression for air resistance, but let us assume it is correct.) Obtain an expression for the integral to be minimized.

2.12 Determine the curve of length l that joins points x_1 and x_2 on the axis, such that the area enclosed by the curve and the x axis is maximized.

2.13 Determine the shape of a cylinder (that is, the ratio of radius to height) that will minimize the surface area for a given volume.

3

Lagrangian dynamics

This chapter is an introduction to Lagrangian dynamics. We begin by considering d'Alembert's principle and derive Lagrange's equations of motion from it. We discuss virtual work in detail. Then we present Hamilton's principle and use it to carry out a second derivation of Lagrange's equations. (This is similar to the derivation of the Euler–Lagrange equation presented in Chapter 2.) It should be noted that Hamilton's principle is the fundamental principle of analytical mechanics. The fact that it is deceptively simple should not obscure the fact that it is very profound. We then consider how to determine the forces of constraint, and finish up by discussing the invariance of Lagrange's equations.

3.1 The principle of d'Alembert. A derivation of Lagrange's equations

Recall that the Lagrange equations of motion are

$$\frac{d}{dt}\frac{\partial L}{\partial \dot{q}_i} - \frac{\partial L}{\partial q_i} = 0, \quad i = 1, 2, \ldots, n. \tag{3.1}$$

In this section we present a simple derivation of Lagrange's equations based on d'Alembert's principle.

This principle is, in a sense, a re-statement of Newton's second law, but it is expressed in a way that makes it a very useful concept in advanced mechanics. For a system of N particles, d'Alembert's principle is

$$\sum_{\alpha=1}^{N} \left(\mathbf{F}_{\alpha}^{ext} - \dot{\mathbf{p}}_{\alpha} \right) \cdot \delta \mathbf{r}_{\alpha} = 0, \tag{3.2}$$

where $\mathbf{F}_{\alpha}^{ext}$ is the external force acting on particle α and $\dot{\mathbf{p}}_{\alpha}$ is the change in momentum of that particle with respect to time. By Newton's second law it is clear that the term in parenthesis is zero. Therefore, multiplying this term

70

by $\delta \mathbf{r}_\alpha$ has no effect. Furthermore, summing over all particles is just a sum of terms all equal to zero. Thus the fact that the expression is equal to zero is obviously true. What is not so obvious (yet) is why this particular formulation is of any value.

We will begin by expressing d'Alembert's principle in Cartesian coordinates and then transforming to generalized coordinates. For N particles there are $n = 3N$ Cartesian coordinates, and we can express d'Alembert's principle in scalar form as

$$\sum_{i=1}^{n} \left(F_i^{ext} - \dot{p}_i \right) \delta x_i = 0. \tag{3.3}$$

If there are n Cartesian coordinates and k constraints, we can (in principle) express the problem in terms of $n - k$ generalized coordinates. The transformation equations are

$$x_1 = x_1(q_1, q_2, \ldots, q_{n-k}, t)$$
$$x_2 = x_2(q_1, q_2, \ldots, q_{n-k}, t)$$
$$\vdots$$
$$x_n = x_n(q_1, q_2, \ldots, q_{n-k}, t).$$

Note that although there are n Cartesian coordinates x_i there are only $n - k$ generalized coordinates because each constraint reduces by one the number of degrees of freedom and hence the number of generalized coordinates.

The quantity $\delta \mathbf{r}_i$ in (3.2) or δx_i in (3.3) is a *virtual displacement*. According to Equation (1.9),

$$\delta x_i = \sum_{j=1}^{n-k} \frac{\partial x_i}{\partial q_j} \delta q_j.$$

Therefore, the first term in d'Alembert's principle can be written

$$\sum_i F_i^{ext} \delta x_i = \sum_i F_i^{ext} \left(\sum_j \frac{\partial x_i}{\partial q_j} \delta q_j \right) = \sum_{i,j} \left(F_i^{ext} \frac{\partial x_i}{\partial q_j} \right) \delta q_j = \sum_j Q_j \delta q_j.$$

The last quantity is the "virtual work." (It is the work done during the virtual displacement. See Section 1.6.)

Next consider the other term in d'Alembert's principle, namely $\dot{p}_i \delta x_i$:

$$\sum_i^n \dot{p}_i \delta x_i = \sum_i^n \dot{p}_i \sum_j^{n-k} \frac{\partial x_i}{\partial q_j} \delta q_j = \sum_i^n m_i \ddot{x}_i \sum_j^{n-k} \frac{\partial x_i}{\partial q_j} \delta q_j,$$

where we assumed the masses of the particles are constant. But,

$$\frac{d}{dt}\left(m_i \dot{x}_i \frac{\partial x_i}{\partial q_j}\right) = m_i \ddot{x}_i \frac{\partial x_i}{\partial q_j} + m_i \dot{x}_i \frac{d}{dt}\frac{\partial x_i}{\partial q_j}.$$

Recall Equation (1.7):

$$\frac{\partial x_i}{\partial q_j} = \frac{\partial \dot{x}_i}{\partial \dot{q}_j},$$

and note further that the last term of the previous equation involves the expression

$$\frac{d}{dt}\frac{\partial x_i}{\partial q_j} = \frac{\partial v_i}{\partial q_j} = \frac{\partial \dot{x}_i}{\partial q_j}.$$

Consequently,

$$\sum_i \dot{p}_i \delta x_i = \sum_{i,j}\left[\frac{d}{dt}\left(m_i \dot{x}_i \frac{\partial x_i}{\partial q_j}\right) - m_i \dot{x}_i \frac{\partial \dot{x}_i}{\partial q_j}\right]\delta q_j,$$

$$= \sum_{i,j}\left[\frac{d}{dt}\left(m_i \dot{x}_i \frac{\partial \dot{x}_i}{\partial \dot{q}_j}\right) - m_i \dot{x}_i \frac{\partial \dot{x}_i}{\partial q_j}\right]\delta q_j.$$

But

$$\frac{\partial T}{\partial q_j} = \frac{\partial}{\partial q_j}\sum_i\left(\frac{1}{2}m_i \dot{x}_i^2\right) = \sum_i m_i\left(\dot{x}_i \frac{\partial \dot{x}_i}{\partial q_j}\right),$$

and

$$\frac{\partial T}{\partial \dot{q}_j} = \frac{\partial}{\partial \dot{q}_j}\sum_i\left(\frac{1}{2}m_i \dot{x}_i^2\right) = \sum_i m_i\left(\dot{x}_i \frac{\partial \dot{x}_i}{\partial \dot{q}_j}\right),$$

so finally,

$$\sum_i \dot{p}_i \delta x_i = \sum_j\left[\frac{d}{dt}\frac{\partial T}{\partial \dot{q}_j} - \frac{\partial T}{\partial q_j}\right]\delta q_j,$$

and d'Alembert's principle reads

$$\sum_{i=1}^n\left(F_i^{ext} - \dot{p}_i\right)\delta x_i = \sum_j Q_j \delta q_j - \sum_j\left[\frac{d}{dt}\frac{\partial T}{\partial \dot{q}_j} - \frac{\partial T}{\partial q_j}\right]\delta q_j = 0,$$

$$0 = \sum_j\left(Q_j - \frac{d}{dt}\frac{\partial T}{\partial \dot{q}_j} + \frac{\partial T}{\partial q_j}\right)\delta q_j.$$

Since the δq_j are independent, each term in the series must individually be equal to zero, so

$$\frac{d}{dt}\frac{\partial T}{\partial \dot{q}_j} - \frac{\partial T}{\partial q_j} = Q_j. \tag{3.4}$$

This is called the Nielsen form of Lagrange's equations. Thus we have derived Lagrange's equations in terms of generalized forces and kinetic energy. If the generalized forces are derivable from a potential that is independent of the generalized velocities, we can write $Q_j = -\partial V/\partial q_j$, and consequently,

$$\frac{d}{dt}\frac{\partial T}{\partial \dot{q}_j} - \frac{\partial T}{\partial q_j} = -\frac{\partial V}{\partial q_j}.$$

That is,

$$\frac{d}{dt}\frac{\partial (T-V)}{\partial \dot{q}_j} - \frac{\partial (T-V)}{\partial q_j} = 0.$$

This yields the familiar form of Lagrange's equations:

$$\frac{d}{dt}\frac{\partial L}{\partial \dot{q}_j} - \frac{\partial L}{\partial q_j} = 0. \tag{3.5}$$

Equations (3.4) and (3.5) are equivalent expressions for Lagrange's equations.

Exercise 3.1 Using the Nielsen form, determine the equation of motion for a mass m connected to a spring of constant k.

Exercise 3.2 Using the Nielsen form, determine the equations of motion for a planet in orbit around the Sun. (Answer: $m\ddot{r} - mr\dot{\theta}^2 = -\frac{GMm}{r^2}$ and $mr\ddot{\theta} + 2m\dot{r}\dot{\theta} = 0$.)

3.2 Hamilton's principle

A mechanical system composed of N particles can be described by $n = 3N$ Cartesian coordinates. All of these coordinates may not be independent. If there are k equations of constraint, the number of independent coordinates is $n - k$. The set of values for all the coordinates at a given instant of time is called the configuration of the system at that time. That is, the configuration is given by $\{q_1, q_2, \ldots, q_{n-k}\}$. We shall assume that all of the generalized coordinates can be varied independently. (When we consider how to determine the forces

of constraint, we will formulate the problem in terms of generalized coordinates that are not all independent. But, for now, keep in mind that all the q_is
are independent.)

Recall that the system can be represented as a point in the $(n - k)$ dimensional space called configuration space. As time goes on, the coordinates of
the particles will change in a continuous manner and the point describing the
system will move in configuration space. As the system goes from an initial
configuration to a final configuration this point traces out a path or trajectory
in configuration space. Obviously there are infinitely many paths between the
initial point and the final point. We know, however, that a mechanical system
always takes a particular path between these two points. What is special about
the path actually taken? Hamilton's principle answers this question. It states
that the path taken by a system is the path that minimizes the "action." The
action of a mechanical system is defined as the line integral

$$I = \int_{t_1}^{t_2} L dt, \tag{3.6}$$

where

$$L = L(q_1, q_2, \ldots, q_{n-k}; \dot{q}_1, \dot{q}_2, \ldots, \dot{q}_{n-k}; t),$$

is the Lagrangian of the system and the limits on the integral, t_1 and t_2, are
the initial and final times for the process. Hamilton's principle tells us that
the system behaves in such a way as to minimize this integral. Therefore, the
variation of the integral is zero. In equation form, Hamilton's principle is

$$\delta \int_{t_1}^{t_2} L dt = 0. \tag{3.7}$$

To interpret this relation, imagine a configuration space and an initial and final
point for some process as indicated in Figure 3.1. The figure shows several

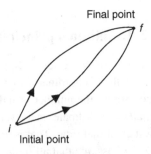

Figure 3.1   Three paths in configuration space.

paths between the end points i and f. Assume that one of these paths is the one that minimizes the action. (It is the "true" path.) Then, along that path, $\delta \int_{t_1}^{t_2} L dt = 0$. The integral is a minimum for the path the system actually follows, and along any other path the value of the action will be greater.

3.3   Derivation of Lagrange's equations

We can now derive Lagrange's equations from Hamilton's principle, recognizing that a simple change of variables from $\Phi(x, y, y')$ to $L(t, q, \dot{q})$ leads to the nearly trivial reformulation of Hamilton's principle into the language of the calculus of variations. Hamilton's principle tells us that a mechanical system will follow a path in configuration space given by $q = q(t)$ such that the integral

$$\int_{t_i}^{t_f} L(t, q, \dot{q}) dt$$

is minimized. (Note that here the Lagrangian is to be considered a functional.) Now if $q = q(t)$ is the minimizing path, then L must satisfy the Euler–Lagrange equation

$$\frac{d}{dt}\left(\frac{\partial L}{\partial \dot{q}}\right) - \frac{\partial L}{\partial q} = 0. \tag{3.8}$$

It is customary to call this equation simply "Lagrange's" equation. It can easily be generalized to systems described by an arbitrary number of generalized coordinates.

Exercise 3.3 Show that Equation (3.7) leads to Equation (3.8). Use the argument given in Section 2.2 as a guide.

3.4   Generalization to many coordinates

Consider a system described by the generalized coordinates $q_1, q_2, \ldots, q_n$. We assume the coordinates are all independent. The Lagrangian for this system is $L(q_1, q_2, \ldots, q_n, \dot{q}_1, \dot{q}_2, \ldots, \dot{q}_n, t)$. Our plan is to show that Lagrange's equations in the form

$$\frac{d}{dt}\left(\frac{\partial L}{\partial \dot{q}_i}\right) - \frac{\partial L}{\partial q_i} = 0, \quad i = 1, \ldots, n$$

follow from Hamilton's principle. The derivation is similar to that presented in Section 2.3.

Recall the definition of the variation of a function. If

$$f = f(q_i, \dot{q}_i, t),$$

then

$$\delta f = \sum_i \left[\left(\frac{\partial f}{\partial q_i} \right) \delta q_i + \left(\frac{\partial f}{\partial \dot{q}_i} \right) \delta \dot{q}_i \right].$$

Note that in this expression, the time is assumed constant.

Consider again Hamilton's principle in the form of Equation (3.7), which is

$$\delta \int_{t_1}^{t_2} L dt = \delta \int_{t_1}^{t_2} L(q_1, q_2, \ldots, q_n, \dot{q}_1, \dot{q}_2, \ldots, \dot{q}_n, t) dt = 0,$$

or

$$\delta I = 0,$$

where the action is represented by I, and the integral is taken over the "path" actually followed by the physical system. Note that for this path the variation in the action is zero. However, as before, there are an infinite number of possible paths between the end points, and we express them in the form

$$Q_i(t) = q_i(t) + \epsilon \eta_i(t).$$

In this case, the action can be considered a function of ϵ. The Lagrangian L is a function of Q_i, $\dot{Q}_i$ and (indirectly) of ϵ. Therefore, the variation of the action is

$$\delta I = \int_{t_1}^{t_2} \delta L dt = \int_{t_1}^{t_2} \sum_i \left(\frac{\partial L}{\partial Q_i} \delta Q_i + \frac{\partial L}{\partial \dot{Q}_i} \delta \dot{Q}_i \right) dt.$$

Noting that Q_i and $\dot{Q}_i$ are functions of ϵ, we write

$$\delta I = \frac{\partial I}{\partial \epsilon} d\epsilon = \int_{t_1}^{t_2} \sum_i \left(\frac{\partial L}{\partial Q_i} \frac{\partial Q_i}{\partial \epsilon} d\epsilon + \frac{\partial L}{\partial \dot{Q}_i} \frac{\partial \dot{Q}_i}{\partial \epsilon} d\epsilon \right) dt.$$

The second term can be integrated by parts. (Note that the mathematical procedure here is nearly identical to that from Equation (2.6) to Equation (2.7).)

We have

$$\int_{t_1}^{t_2} \frac{\partial L}{\partial \dot{Q}_i} \frac{\partial \dot{Q}_i}{\partial \epsilon} d\epsilon dt = \int_{t_1}^{t_2} \frac{\partial L}{\partial \dot{Q}_i} \frac{\partial^2 Q_i}{\partial \epsilon \partial t} dt d\epsilon = \int_{t_1}^{t_2} \frac{\partial L}{\partial \dot{Q}_i} \frac{\partial}{\partial \epsilon} \left(\frac{\partial Q_i}{\partial t} dt \right) d\epsilon$$

$$= \frac{\partial L}{\partial \dot{Q}_i} \frac{\partial Q_i}{\partial \epsilon} \Big|_{t_1}^{t_2} - \int_{t_1}^{t_2} \frac{\partial Q_i}{\partial \epsilon} \frac{d}{dt} \frac{\partial L}{\partial \dot{Q}_i} dt d\epsilon$$

$$= - \int_{t_1}^{t_2} \frac{\partial Q_i}{\partial \epsilon} \frac{d}{dt} \frac{\partial L}{\partial \dot{Q}_i} dt d\epsilon,$$

where we have used the fact that $\frac{\partial Q_i}{\partial \epsilon} \Big|_{t_1}^{t_2} = 0$ because all the curves pass through the end points.

Consequently,

$$\delta I = \int_{t_1}^{t_2} \sum_i \left(\frac{\partial L}{\partial Q_i} \frac{\partial Q_i}{\partial \epsilon} - \frac{\partial Q_i}{\partial \epsilon} \frac{d}{dt} \frac{\partial L}{\partial \dot{Q}_i} \right) d\epsilon dt$$

$$= \int_{t_1}^{t_2} \sum_i \left(\frac{\partial L}{\partial Q_i} - \frac{d}{dt} \frac{\partial L}{\partial \dot{Q}_i} \right) \frac{\partial Q_i}{\partial \epsilon} d\epsilon dt$$

$$= \int_{t_1}^{t_2} \sum_i \left(\frac{\partial L}{\partial Q_i} - \frac{d}{dt} \frac{\partial L}{\partial \dot{Q}_i} \right) \delta Q_i dt.$$

When $\epsilon = 0$, $\delta I = 0$ because I is minimized along the path $q_i = q_i(t)$. So

$$[\delta I]_{\epsilon=0} = 0 = \int_{t_1}^{t_2} \sum_i \left(\frac{\partial L}{\partial q_i} - \frac{d}{dt} \frac{\partial L}{\partial \dot{q}_i} \right) \delta q_i dt. \qquad (3.9)$$

Since the δq_is are independent, the only way this expression can be zero is for all the terms in the parenthesis to be zero. Hence, Lagrange's equations are proved.

3.5 Constraints and Lagrange's λ-method

We now apply Lagrange's λ-method of Section 2.4.1 to determine the forces of constraint acting on a system.

Let a system be described by n generalized coordinates $q_1, q_2, \ldots, q_n$. Assume, further, that the system is subjected to k holonomic constraints, so there are k equations relating the coordinates. Obviously, not all of the generalized coordinates are independent, so we could use the k equations of constraint to reduce the number of generalized coordinates to $n - k$. However, we prefer, for the moment, to use all n of the coordinates.

The behavior of the system is described by Hamilton's principle,

$$\delta I = \delta \int_{t_1}^{t_2} L dt = 0,$$

and the k equations of constraint

$$f_j(q_1, q_2, \ldots, q_n) = 0, \qquad j = 1, \ldots, k. \tag{3.10}$$

If all the coordinates were independent, Hamilton's principle would lead to n equations of the form

$$\frac{d}{dt}\left(\frac{\partial L}{\partial \dot{q}_i}\right) - \frac{\partial L}{\partial q_i} = 0, \quad i = 1, \ldots, n. \tag{3.11}$$

But we cannot write this (yet) because as noted following Equation (3.9), this result depends on all the δqs being independent.

Equations (3.10) represent *holonomic* constraints. Consequently

$$\delta f_j = \frac{\partial f_j}{\partial q_1}\delta q_1 + \frac{\partial f_j}{\partial q_2}\delta q_2 + \cdots + \frac{\partial f_j}{\partial q_n}\delta q_n = 0.$$

That is,

$$\sum_{i=1}^{n} \frac{\partial f_j}{\partial q_i}\delta q_i = 0 \qquad j = 1, \ldots, k. \tag{3.12}$$

Here the coefficients $\partial f_j/\partial q_i$ are functions of the q_i. If Equation (3.12) holds, then multiplying it by some quantity, call it λ_j, will have no effect, so we can write it as

$$\lambda_j \sum_{i=1}^{n} \frac{\partial f_j}{\partial q_i}\delta q_i = 0.$$

The λs are undetermined multipliers. There are k such equations and adding them together still yields zero, so

$$\sum_{j=1}^{k}\lambda_j \sum_{i=1}^{n} \frac{\partial f_j}{\partial q_i}\delta q_i = 0.$$

Further, we can integrate from t_i to t_f and still not change the fact that the value is zero. This yields the useful relation

$$\int_{t_1}^{t_2} dt \left(\sum_{j=1}^{k}\lambda_j \sum_{i=1}^{n} \frac{\partial f_j}{\partial q_i}\delta q_i\right) = 0. \tag{3.13}$$

Equation (3.9) can be written as

$$\int_{t_1}^{t_2} dt \sum_i \left(\frac{\partial L}{\partial q_i} - \frac{d}{dt} \frac{\partial L}{\partial \dot{q}_i} \right) \delta q_i = 0. \tag{3.14}$$

Adding Equations (3.13) and (3.14) we obtain

$$\int_{t_1}^{t_2} dt \sum_i \left(\frac{\partial L}{\partial q_i} - \frac{d}{dt} \frac{\partial L}{\partial \dot{q}_i} + \sum_{j=1}^{k} \lambda_j \frac{\partial f_j}{\partial q_i} \right) \delta q_i = 0. \tag{3.15}$$

The δqs are not all independent; they are related by the k equations of constraint. However, $n - k$ of them are independent and for these coordinates the term in parenthesis in Equation (3.15) must be zero. For the remaining k equations we can *select* the undetermined multipliers λ_j such that

$$\frac{\partial L}{\partial q_i} - \frac{d}{dt} \frac{\partial L}{\partial \dot{q}_i} + \sum_{j=1}^{k} \lambda_j \frac{\partial f_j}{\partial q_i} = 0.$$

Consequently, we have, for all q_is, the following relations

$$\frac{d}{dt} \frac{\partial L}{\partial \dot{q}_i} - \frac{\partial L}{\partial q_i} = \sum_{j=1}^{k} \lambda_j \frac{\partial f_j}{\partial q_i}, \qquad i = 1, \dots, n. \tag{3.16}$$

Equations (3.16) are n equations and Equations (3.10) give us k equations so we have $n + k$ equations which can be solved to obtain expressions for the qs and for the λs.

We now show how the λs are related to the generalized forces of constraint. Consider Lagrange's equations written in the Nielsen form (Equation (3.4)). Let us write the equation in the following way, in which we separate the conservative forces Q_i^c derivable from a potential function (V) from the forces of constraint Q_i^{nc} (which in general cannot be derived from a potential that depends only on the coordinates):

$$\frac{d}{dt} \frac{\partial T}{\partial \dot{q}_i} - \frac{\partial T}{\partial q_i} = Q_i^c + Q_i^{nc}.$$

The conservative force (Q^c) can be expressed as $-\frac{\partial V}{\partial q_i}$. Assuming that $\partial V / \partial \dot{q}_i = 0$, we can write this equation as

$$\frac{d}{dt} \frac{\partial L}{\partial \dot{q}_i} - \frac{\partial L}{\partial q_i} = Q_i^{nc}. \tag{3.17}$$

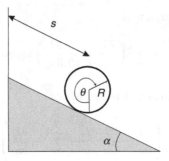

Figure 3.2   A disk rolling without slipping on an inclined plane.

But Equations (3.16) and (3.17) must be identical. Hence we have the following expression

$$Q_i^{nc} = \sum_{j=1}^{k} \lambda_j \frac{\partial f_j}{\partial q_i},$$

and we have obtained the desired expression for the forces of constraint.

Example 3.1 *Consider a disk of radius R rolling down an inclined plane of length l and angle α. Find the equations of motion, the angular acceleration, and the force of constraint. See Figure 3.2.*

Solution 3.1 *The moment of inertia of a disk is $I = \frac{1}{2}MR^2$, and the kinetic energy is $T = \frac{1}{2}M\dot{s}^2 + \frac{1}{2}I\dot{\theta}^2$. The potential energy is $V = Mg(l - s)\sin\alpha$. Therefore,*

$$L = \frac{1}{2}M\dot{s}^2 + \frac{1}{4}MR^2\dot{\theta}^2 + Mg(s - l)\sin\alpha.$$

The disk is constrained to roll without slipping. Therefore,

$$f(s, \theta) = s - R\theta = 0.$$

This is a holonomic constraint. Consequently,

$$\frac{d}{dt}\frac{\partial L}{\partial \dot{s}} - \frac{\partial L}{\partial s} = \lambda \frac{\partial f}{\partial s} \quad \text{where} \quad \frac{\partial f}{\partial s} = 1,$$

$$\frac{d}{dt}\frac{\partial L}{\partial \dot{\theta}} - \frac{\partial L}{\partial \theta} = \lambda \frac{\partial f}{\partial \theta} \quad \text{where} \quad \frac{\partial f}{\partial \theta} = -R.$$

Carrying out the indicated operations we obtain the two equations of motion

$$\frac{d}{dt}(M\dot{s}) - Mg\sin\alpha = \lambda,$$

$$\frac{d}{dt}\left(\frac{1}{2}MR^2\dot{\theta}\right) = -R\lambda.$$

Furthermore, from the constraint equation we obtain one more equation, namely,

$$\ddot{\theta} = \ddot{s}/R.$$

You can easily show that

$$\ddot{\theta} = \frac{2}{3}\frac{g\sin\alpha}{R},$$

and

$$\ddot{s} = \frac{2}{3}g\sin\alpha.$$

It follows that

$$\lambda = M\ddot{s} - Mg\sin\alpha = -\frac{1}{3}Mg\sin\alpha.$$

Now the forces of constraint are given by

$$Q_s = \lambda\frac{\partial f}{\partial s} = -\frac{1}{3}Mg\sin\alpha = tangential\ force,$$

and

$$Q_\theta = \lambda\frac{\partial f}{\partial\theta} = \frac{1}{3}MgR\sin\alpha = torque.$$

Q_s *and* Q_θ *are the generalized forces required to keep the disk rolling down the plane without slipping.*

Exercise 3.4 Fill in the missing steps in the example above to show that $\ddot{\theta} = (2/3)g\sin\alpha/R$ and $\ddot{s} = (2/3)g\sin\alpha$.

3.6 Non-holonomic constraints

Certain non-holonomic constraints can also be treated by Lagrange's λ-method, as described in Section 2.4.2. Suppose the non-holonomic constraints are expressed as relations among the differentials of the coordinates. For example, if there were only one such constraint, it would be given by

$$A_1 dq_1 + A_2 dq_2 + \cdots + A_n dq_n = 0,$$

where the As are known functions of the qs. If this relation holds for the dqs, it will also hold for the δqs, so we can write

$$A_1 \delta q_1 + A_2 \delta q_2 + \cdots + A_n \delta q_n = 0. \tag{3.18}$$

But this last expression has the same form as the expression for δf for a holonomic constraint, which we wrote as

$$\delta f = \frac{\partial f}{\partial q_1} \delta q_1 + \frac{\partial f}{\partial q_2} \delta q_2 + \cdots + \frac{\partial f}{\partial q_n} \delta q_n = 0.$$

Therefore, we can use exactly the same reasoning as before, simply replacing the $\frac{\partial f}{\partial q}$s with As. (Note, however, that although the As are known quantities, they are not partial derivatives of some known function.) Consequently, if our single non-holonomic constraint can be expressed as in Equation (3.18), the Lagrange equations are

$$\frac{d}{dt} \frac{\partial L}{\partial \dot{q}_i} - \frac{\partial L}{\partial q_i} = \lambda A_i, \quad i = 1, 2, \ldots, n.$$

If there is more than one non-holonomic constraint, we need to generalize this relation. For example, if there are m non-holonomic constraints we will have i equations of the form

$$\frac{d}{dt} \frac{\partial L}{\partial \dot{q}_i} - \frac{\partial L}{\partial q_i} = \sum_{k=1}^{m} \lambda_k A_{ki} \quad i = 1, 2, \ldots, n.$$

We can even use this technique for time dependent ("rheonomic") constraints having the form

$$\sum_{i=1}^{n} A_{ki} dq_i + B_{kt} dt = 0, \quad k = 1, 2, \ldots, m.$$

Here the coefficients (As and Bs) are, in general, functions of both the coordinates and time. As before, we can replace d with δ and write these constraints in the form

$$\sum_{i=1}^{n} A_{ki} \delta q_i + B_{kt} \delta t = 0, \quad k = 1, 2, \ldots, m.$$

The coefficients of δt (the Bs) do not enter into the equation of motion because time is frozen during the virtual displacements, δq_i. Thus the equations of motion take on the familiar form

$$\frac{d}{dt}\frac{\partial L}{\partial \dot{q}_i} - \frac{\partial L}{\partial q_i} = \sum_{k=1}^{m} \lambda_k A_{ki}, \quad i = 1, 2, \ldots, n.$$

Note, however, that the Bs enter into the relations between the velocities, thus

$$A_{i1}\dot{q}_1 + A_{i2}\dot{q}_2 + \cdots + A_{in}\dot{q}_n + B_i = 0, \quad i = 1, 2, \ldots, n.$$

3.7 Virtual work

We now consider the concept of virtual work in more detail than was done in Section 1.6. A mechanical system will be in equilibrium if and only if the total virtual work of all the "impressed" forces is zero. (Impressed forces are those applied to the system and do not include forces of constraint.[1]) The principle of virtual work for a system in equilibrium is expressed mathematically as

$$\delta W = \sum_j Q_j \delta q_j = 0.$$

Here the virtual displacements δq_i satisfy all the constraints. Note that the principle of virtual work is a variational principle.

It is interesting to contrast the principle of virtual work with the Newtonian concept of equilibrium. In Newtonian mechanics when a system is in equilibrium, the vector sum of *all* of the forces acting on it is zero. That is, Newtonian mechanics replaces the constraints by forces. Analytical mechanics, on the other hand, does away with the reaction forces and only considers the impressed forces; however, it requires that any virtual displacements satisfy the constraints on the system. (As we shall see, considering virtual displacements that violate the constraints allows us to determine the forces of constraint.)

In vector notation the generalized forces are just the components of the actual forces and the principle of virtual work can be expressed as

$$\mathbf{F}_1 \cdot \delta \mathbf{r}_1 + \mathbf{F}_2 \cdot \delta \mathbf{r}_2 + \cdots + \mathbf{F}_N \cdot \delta \mathbf{r}_N = 0.$$

In terms of Cartesian coordinates, this is

$$F_{1x}\delta x_1 + F_{1y}\delta y_1 + \cdots + F_{Nz}\delta z_N = 0.$$

---

[1] The forces of constraint are often referred to as "reaction" forces.

When written in this form, we appreciate that the scalar products of the forces
and the virtual displacements are all zero. This means that the force $\mathbf{F}_i$ is
perpendicular to any allowed displacement of particle i. For a free particle,
all displacements are allowed, so the net force must be zero. For a particle
restricted to a table top, the force must be perpendicular to the table top.

If an equilibrium problem involves constraints, then the equilibrium condition can be determined using the Lagrange λ-method.

We have justified the principle of virtual work, but we have not proved it.
Let us leave it as a postulate. (In fact, it is the only postulate of analytical
mechanics and is applicable to dynamical as well as equilibrium situations.[2])

Postulate   The virtual work for a system in equilibrium is zero,

$$\delta W = 0.$$

Since Newtonian dynamics states that the total force on a system in equilibrium is zero, we conclude that the reaction forces are equal and opposite to the
impressed forces. Thus, the postulate can be stated in terms of reaction forces
as follows.

Postulate   The virtual work of any force of reaction is zero for any virtual
displacement that satisfies the constraints on the system.

Exercise 3.5 Assume the impressed force is derivable from a scalar function. Show that the equilibrium state is given by the stationary value of the
potential energy, $\delta V = 0$.

3.7.1 Physical interpretation of the Lagrange multipliers

In equilibrium, the virtual work of the impressed forces is zero:

$$\delta W = 0.$$

If the impressed forces can be derived from a potential energy ($F_i = -\partial V/\partial q_i$)
then the virtual work is equal to the negative of the variation of the potential
energy because

$$\delta W = \Sigma_i F_i \delta q_i = -\Sigma_i (\partial V/\partial q_i)\delta q_i = -\delta V.$$

If the physical system is subjected to a *holonomic* constraint given by

$$f(q_1, \ldots, q_n) = 0,$$

---

[2] For a discussion of the postulate, see Lanczos, *op. cit.* page 76.

then the Lagrange λ-method requires that

$$\delta(L + \lambda f) = 0.$$

(See Equation (2.24).) Since λ is undetermined, we can just as well use $-\lambda$ and define a modified Lagrangian by

$$\overline{L} = L - \lambda f.$$

Since $L = T - V$, this is equivalent to defining a modified potential energy $V + \lambda f$. Therefore, $\delta W = 0$ is equivalent to

$$\delta(V + \lambda f) = 0.$$

If we permit arbitrary variations of the generalized coordinates (not just those that satisfy the constraint), then reaction forces *will* act on the system. Consequently, the term λf can be interpreted as the potential energy related to the force of constraint. Thus, the x_i component of the reaction force is

$$F_i = -\frac{\partial(\lambda f)}{\partial x_i} = -\lambda \frac{\partial f}{\partial x_i} - \frac{\partial \lambda}{\partial x_i} f.$$

But $f = 0$, so

$$F_i = -\lambda \frac{\partial f}{\partial x_i}.$$

We conclude that a holonomic constraint is maintained by forces that can be derived from a scalar function.

As we have seen, some non-holonomic constraints can also be dealt with using the Lagrangian λ-method; however, these forces (friction is an example) cannot be derived from a scalar function.

Example 3.2 *You have probably proved by using calculus (perhaps in a previous mechanics course) that the curve formed by a hanging chain or heavy rope is a catenary. That problem can also be solved using variational techniques. The quantity to be minimized is the potential energy, $V = V(y)$. The problem is to determine $y = y(x)$ subject to the constraint that the length of the chain is a given constant. In variational terms, we want to determine $y = y(x)$ given that $\delta V = 0$ and subject to the constraint $\int_{x_1}^{x_2} ds = constant = length = l.$*

Solution 3.2 *The constraint can be expressed as*

$$l = \int_{x_1}^{x_2} \sqrt{1 + y'^2} dx.$$

The potential energy of a length of chain ds is $dV = (dm)gh = \rho g y ds$. *Dropping unnecessary constants,*

$$V = \int_{x_1}^{x_2} y ds = \int_{x_1}^{x_2} y \sqrt{1 + y'^2} dx.$$

The Lagrange λ-method leads to

$$\delta \int_{x_1}^{x_2} (y + \lambda) \sqrt{1 + y'^2} dx = 0.$$

Determining the equation of the curve is left as an exercises.

Exercise 3.6 The functional in the preceding example does not depend on x. Reformulate the problem as discussed in Section 2.2.2 and show that the curve is a catenary.

Exercise 3.7 Consider a simple pendulum composed of a mass m constrained by a wire of length l to swing in an arc. Assume r and θ are both variables. Obtain the Lagrange equations and using the λ-method, determine the tension in the wire. Check by using elementary methods.

3.8 The invariance of the Lagrange equations

The Lagrange equations are invariant under a point transformation. A point transformation is a transformation from a set of coordinates $q_1, q_2, \ldots, q_n$ to a set $s_1, s_2, \ldots, s_n$ such that for every point P_q in "q-space" there corresponds a point P_s in "s-space." The transformation equations have the form

$$q_1 = q_1(s_1, s_2, \ldots, s_n, t),$$

$$\vdots$$

$$q_n = q_n(s_1, s_2, \ldots, s_n, t).$$

It is not difficult to show that the Lagrange equations are invariant under a point transformation. (That is, they retain the same form. See Problem 3.3 at the end of the chapter.) However, it may be somewhat more difficult to appreciate the significance of this invariance, so let us consider it in more detail.

Figure 3.3 shows an n-dimensional configuration space. At a given instant, the system is represented by a single point. As time goes on, this point moves

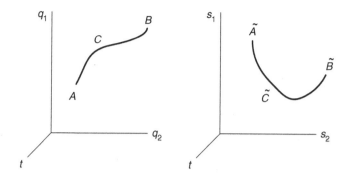

Figure 3.3   "True" path in two different coordinate systems.

along a curve.[3] The system moves from end point A to end point B along the curve C that minimizes the action. By considering varied paths between the same end points and requiring that the action integral be stationary for the "true" path, we obtain the conditions

$$\frac{d}{dt}\frac{\partial L}{\partial \dot{q}_i} - \frac{\partial L}{\partial q_i} = 0, \qquad i = 1, \ldots, n.$$

Let the coordinates q_i be transformed to s_i by a point transformation. The end points are transformed to $\widetilde{A}$, $\widetilde{B}$ and the curve C is transformed to $\widetilde{C}$, as shown in the right-hand panel of Figure 3.3.

The action integral is minimized by the "true" path between the end points in any coordinate system, consequently the Lagrange equations are valid in the new coordinate system. Note, however, that although the *value* of L is the same in both coordinate system, its *form* may be quite different. *The invariance of the Lagrange equations allows us to select whatever set of coordinates makes the equations easiest to solve.* (The invariance principle becomes a postulate in the theory of special relativity in which Einstein postulated that the laws of physics are invariant under transformations between inertial reference frames.)

The Lagrangian equation of motion for a particular i will not, in general, be the same in the two systems, but the complete sets of equations are equivalent. For this reason, one should actually refer to the equations as co-variant rather than in-variant. For example, if

$$L = \frac{1}{2}\dot{x}^2 + \frac{1}{2}\dot{y}^2 + \frac{1}{(x^2 + y^2)^{1/2}},$$

---

[3] We can include the time as another coordinate, and represent the system as a point in an $(n+1)$-dimensional configuration space. In this space, the system traces out a curve or "world line" that gives the history of the system.

a point transformation, $x = r\cos\theta$, $y = r\sin\theta$, yields the Lagrangian

$$L = \frac{1}{2}\dot{r}^2 + \frac{1}{2}r^2\dot{\theta}^2 + \frac{1}{r}.$$

The equations of motion in terms of x and y are

$$\ddot{x} = -\frac{x}{(x^2 + y^2)^3} \qquad\qquad \ddot{y} = -\frac{y}{(x^2 + y^2)^3},$$

whereas the equations of motion in terms of r and θ are

$$\ddot{r} - r\dot{\theta}^2 + \frac{1}{r} = 0 \qquad\qquad \frac{d}{dt}(r^2\dot{\theta}) = 0.$$

Although these equations of motion are quite different in form, they yield the same result and for a particular set of the position and velocity parameters they yield the same numerical value for the Lagrangian.

3.9 Problems

3.1 Assume the generalized force is derivable from a potential independent of the generalized velocities. Show that Equations (3.1) follow from Equations (3.4).

3.2 Use d'Alembert's principle to determine the equilibrium condition for the system illustrated in Figure 3.4.

3.3 If one expresses the Lagrangian in terms of some set of generalized coordinates, q_i, the Lagrange equations of motion have the form

$$\frac{d}{dt}\frac{\partial L}{\partial \dot{q}_i} - \frac{\partial L}{\partial q_i} = 0.$$

Assume that we transform from the q_i to a new set of coordinates s_i by a so-called "point transformation"

$$q_i = q_i(s_1, s_2, \ldots, s_n, t).$$

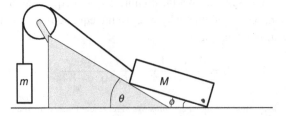

Figure 3.4   A system in equilibrium.

Prove that Lagrange's equations are unchanged. (That is, prove that Lagrange's equations are invariant under a point transformation.)

3.4 Consider an object thrown upward in a constant gravitational field. Demonstrate numerically that $\int_{t_1}^{t_2}(T - V)dt$ is smaller for the true path than for a "false" path. Select any "false" path you wish.

3.5 A rope is looped over two frictionless pegs that are at a height h above the surface of the Earth and are separated horizontally by a distance d. The ends of the rope lie on the surface. Determine the equation of the curve of the hanging portion of the rope.

3.6 A particle slides on the outer surface of an inverted hemisphere. Using Lagrangian multipliers, determine the reaction force on the particle. Where does the particle leave the hemispherical surface?

3.7 A right triangular wedge of angle θ and mass M can slide on a frictionless horizontal surface. A block of mass m is placed on the wedge and it too can slide on the frictionless inclined surface of the wedge. Use the method of Lagrange multipliers and determine the equations of motion for the wedge and the block. Determine the forces of constraint.

3.8 A particle of mass m_1 is connected by an inextensible, massless string to a particle of mass m_2. Particle m_1 is free to slide on the surface of frictionless table. The string passes through a hole in the table and m_2 is hanging freely under the table and allowed to move only in the vertical direction. This arrangement will be stable only if m_1 is moving in a path around the hole. Write the Lagrangian for the system. Solve the equations of motion.

3.9 A free body can rotate about any axis. That is, $\boldsymbol{\Omega}$ can have any arbitrary direction. Consider an allowed displacement of the kth point in the body,

$$\delta \mathbf{r}_k = \epsilon \boldsymbol{\Omega} \times \mathbf{r}_k.$$

Using the principle of virtual work, show that this implies the conservation of total angular momentum.

3.10 According to quantum mechanics, a particle of mass m in a rectangular box with sides a, b, c has energy

$$E = \frac{h^2}{8m}\left(\frac{1}{a^2} + \frac{1}{b^2} + \frac{1}{c^2}\right).$$

Assuming the box has constant volume, show that the energy is minimized if the box is a cube ($a = b = c$).

3.11 Consider a cable of fixed length, suspended from two fixed points. Show that the shape that minimizes the potential energy is a hyperbolic cosine.

PART II

Hamiltonian mechanics

4

Hamilton's equations

In this chapter we consider a radically different formulation of the dynamical problem. We define the Hamiltonian and derive Hamilton's "canonical" equations. These are derived in two different ways, first by using a Legendre transformation on the Lagrangian and secondly by using the stationary property of the action integral. Hamilton's approach gives us a whole new way of looking at mechanics problems. Although Hamilton's approach is often not as convenient as Lagrange's method for solving practical problems, it is, nevertheless, a far superior tool for theoretical studies. Some of the methods developed in Hamiltonian mechanics carry over directly into quantum mechanics, statistical mechanics, and other fields of physics.

4.1 The Legendre transformation

Let us begin by considering the Legendre transformation simply as a mathematical technique. Then (in Section 4.2) we will apply it to the Lagrangian.

The Legendre transformation is useful in a number of branches of physics and you are probably familiar with it from your thermodynamics course. This transformation allows us to take a function of one set of variables, such as

$$f = f(u_1, u_2, \ldots, u_n),$$

and generate a new function in terms of the variables v_i which are defined by

$$v_i = \frac{\partial f}{\partial u_i}. \tag{4.1}$$

We shall denote the new function by g, noting that

$$g = g(v_1, v_2, \ldots, v_n).$$

The Legendre transformation of f (a function of the us) to g (a function of the vs) is accomplished by defining g as follows:

$$g = \sum_{i=1}^{n} u_i v_i - f. \tag{4.2}$$

At first glance you might think that g is a function of both us and vs, but you can appreciate that g is a function only of the vs by taking the differential of its definition. That is,

$$dg = \sum_{i=1}^{n} (u_i dv_i + v_i du_i) - \sum_{i=1}^{n} \frac{\partial f}{\partial u_i} du_i,$$

$$= \sum_{i=1}^{n} u_i dv_i + \sum_{i=1}^{n} \left(v_i - \frac{\partial f}{\partial u_i} \right) du_i.$$

But by the definition of the v_is (Equation (4.1)), the coefficients of the du_is are zero, so

$$dg = \sum_{i=1}^{n} u_i dv_i, \tag{4.3}$$

and consequently g is a function only of the vs. But if dg depends only on the vs, its differential is given by

$$dg = \frac{\partial g}{\partial v_1} dv_1 + \frac{\partial g}{\partial v_2} dv_2 + \cdots + \frac{\partial g}{\partial v_n} dv_n = \sum_{i=1}^{n} \frac{\partial g}{\partial v_i} dv_i. \tag{4.4}$$

Comparing Equations (4.3) and (4.4) we appreciate that

$$u_i = \frac{\partial g}{\partial v_i}. \tag{4.5}$$

Note the symmetry between Equations (4.1) and (4.5). Also note that the two functions f and g can be defined in terms of each other in a symmetrical fashion. That is,

$$g = \sum_{i=1}^{n} u_i v_i - f, \quad \text{and} \quad f = \sum_{i=1}^{n} u_i v_i - g. \tag{4.6}$$

Essentially, that is all there is to Legendre transformations. However, it is worth mentioning one more fact. Suppose that f also depends on some other variables, say $w_1, w_2, \ldots, w_m$, that are independent of the us. Suppose, furthermore, that the ws do not participate in the transformation. We call the us the "active" variables and the ws the "passive" variables. We have

$$f = f(u_1, u_2, \ldots, u_n; w_1, w_2, \ldots, w_m).$$

Defining the vs and g as before,

$$v_i = \frac{\partial f}{\partial u_i},$$

$$g = \sum_{i=1}^{n} u_i v_i - f(u, w).$$

The differential of g is

$$dg = \sum_{i=1}^{n} (u_i dv_i + v_i du_i) - df,$$

$$= \sum_{i=1}^{n} (u_i dv_i + v_i du_i) - \sum_{i=1}^{n} \frac{\partial f}{\partial u_i} du_i - \sum_{i=1}^{m} \frac{\partial f}{\partial w_i} dw_i,$$

$$= \sum_{i=1}^{n} u_i dv_i + \sum_{i=1}^{n} \left(v_i - \frac{\partial f}{\partial u_i} \right) du_i - \sum_{i=1}^{m} \frac{\partial f}{\partial w_i} dw_i.$$

As before, the middle term on the right is zero, leaving us with

$$dg = \sum_{i=1}^{n} u_i dv_i - \sum_{i=1}^{m} \frac{\partial f}{\partial w_i} dw_i,$$

which indicates that g is a function of the vs *and* the ws. But if $g = g(v, w)$, its differential has the form

$$dg = \sum_{i=1}^{n} \frac{\partial g}{\partial v_i} dv_i + \sum_{i=1}^{m} \frac{\partial g}{\partial w_i} dw_i,$$

so comparing the two expressions for dg we find that

$$\frac{\partial g}{\partial w_i} = -\frac{\partial f}{\partial w_i}. \tag{4.7}$$

We shall soon be using this relation involving the passive variables.

Exercise 4.1 Show that the transformation $g = f - \sum_{i=1}^{n} u_i v_i$ is also an acceptable Legendre transformation, in the sense that g is a function only of the vs. (This is the form used in thermodynamics.)

4.1.1 Application to thermodynamics

The Legendre transformation is particularly useful in thermodynamics.[1] Without going into the definitions of the various "thermodynamic potentials"

---

[1] This section may be skipped without loss of continuity.

let us recall that the internal energy U is a function of entropy (S) and volume
(V). That is, $U = U(S, V)$. In thermodynamics we may be interested in considering the change in U when there is a change in some variable, or perhaps
we are interested in determining the conditions under which U remains constant. For example, we might be considering a system undergoing a constant
pressure process. In general both S and V will change, and it may be difficult
to determine the effect on U. In that case it is probably more convenient to
consider the behavior of the enthalpy (H) which is a function of entropy and
pressure, $H = H(S, P)$. If P is constant, the enthalpy is a function of only
one variable. Similarly, for a constant temperature process, we might prefer to
consider the Helmholtz free energy $F = F(T, V)$. If both the temperature and
pressure are constant (as in a chemistry laboratory), we might prefer to consider the Gibbs free energy $G = G(T, P)$. These thermodynamic potentials
(U, H, F, G) are different functions of different variables, yet they are related
because they describe different aspects of the same thermodynamic system. As
you may recall from your thermodynamics course, they are obtained from one
another through Legendre transformations. For example, the familiar first law
of thermodynamics "change in energy = heat into the system − work done by
the system" is expressed by

$$dU = TdS - PdV.$$

But since $U = U(S, V)$

$$dU = \frac{\partial U}{\partial S}dS + \frac{\partial U}{\partial V}dV,$$

so we appreciate that $T = \partial U/\partial S$ and $P = -\partial U/\partial V$. Using these facts we
can generate the other thermodynamic potentials by Legendre transformations.
For example, F is a function depending on T and V. We generate F by simply
using the prescription of Exercise 4.1,

$$F = U - TS.$$

To show that F is indeed a function of T and V we note that

$$\begin{aligned} dF &= dU - TdS - SdT \\ &= TdS - PdV - TdS - SdT \\ &= -PdV - SdT. \end{aligned}$$

Therefore $F = U - TS$ is indeed a function of T and V.

Exercise 4.2 In the transformation described above, from internal energy to Helmholtz free energy, identify the active variables and the passive variables.

Exercise 4.3 Given $dU = TdS - PdV$, determine $H(S, P)$ and $G(T, P)$. Answers: $H = U + PV$, $G = U - TS + PV$.

4.2 Application to the Lagrangian. The Hamiltonian

We now apply the Legendre transformation to the Lagrangian. The Lagrangian is a function of $q_i, \dot{q}_i, t$

$$L = L(q_1, \ldots, q_n; \dot{q}_1, \ldots, \dot{q}_n; t).$$

In our Legendre transformation we shall let the $\dot{q}$s be the active variables and the qs and t be the passive variables. Therefore, the new function will depend on the qs and t and on the derivatives of L with respect to the $\dot{q}$s. If the new function is denoted by H we have

$$H = H(q_i, \partial L/\partial \dot{q}_i, t).$$

But we know that $\partial L/\partial \dot{q}_i = p_i$, the generalized momentum conjugate to q_i. Therefore, the new function will be

$$H = H(q_i, p_i, t).$$

We define H by the usual prescription for a Legendre transformation (Equation (4.2)), obtaining

$$H(q_i, p_i, t) = \sum_{i=1}^{n} p_i \dot{q}_i - L(q_i, \dot{q}_i, t). \tag{4.8}$$

The function H is called the Hamiltonian. We shall see shortly that in many situations it is equal to the total energy.

Keep in mind that the Hamiltonian *must* be expressed in terms of the generalized momentum. An expression for H involving velocities is *wrong*. You may consider Equation (4.8) as a definition of the Hamiltonian. (It would be a good idea to memorize it.)

Exercise 4.4 Determine H for a free particle using Equation (4.8). (Answer: $H = (p_x^2 + p_y^2 + p_z^2)/2m$.)

Exercise 4.5 The Lagrangian for a disk rolling down an inclined plane is

$$L = \frac{1}{2}m\dot{y}^2 + \frac{1}{4}mR^2\dot{\theta}^2 + mg(y - l)\sin\alpha.$$

(See Example 3.1.) What is the Hamiltonian for this system? (Answer: $H = p_y^2/2m + p_\theta^2/mR^2 - Mg(y - R)\sin\alpha$.)

4.3 Hamilton's canonical equations

The Hamiltonian is the function of q_i, p_i, t that was obtained from the Lagrangian by assuming that the $\dot{q}_i$s are the active variables and that the q_is and t are the passive variables. Therefore, expressing Equations (4.1) and (4.5) in terms of the variables introduced in the previous section we have

$$p_i = \frac{\partial L}{\partial \dot{q}_i} \quad \text{and} \quad \dot{q}_i = \frac{\partial H}{\partial p_i}. \tag{4.9}$$

Furthermore, from the relations among the passive variables given in Equations (4.7) we have

$$\frac{\partial H}{\partial q_i} = -\frac{\partial L}{\partial q_i} \quad \text{and} \quad \frac{\partial H}{\partial t} = -\frac{\partial L}{\partial t}. \tag{4.10}$$

Now the Lagrange equation tells us that

$$\frac{d}{dt}\left(\frac{\partial L}{\partial \dot{q}_i}\right) = \frac{\partial L}{\partial q_i}.$$

That is,

$$\dot{p}_i = \frac{\partial L}{\partial q_i}.$$

Therefore, substituting for $\partial L/\partial q_i$ from the first of Equations (4.10) we have

$$\dot{p}_i = -\frac{\partial H}{\partial q_i}. \tag{4.11}$$

This equation and the second equation of (4.9) give us the following two extremely important relations:

$$\dot{q}_i = \frac{\partial H}{\partial p_i}, \tag{4.12}$$

$$\dot{p}_i = -\frac{\partial H}{\partial q_i}. \tag{4.13}$$

These are called "Hamilton's canonical equations." They are the *equations of motion* of the system expressed as $2n$ first-order differential equations. They have the very nice property that the derivatives with respect to time are isolated on the left-hand sides of the equation.

I will now present yet another way to obtain the canonical equations. We begin with the definition

$$H(q_i, p_i, t) = \sum_{i=1}^{n} p_i \dot{q}_i - L(q_i, \dot{q}_i, t). \tag{4.14}$$

Evaluating the partial derivative of the Hamiltonian with respect to the generalized coordinate q_i yields

$$\frac{\partial H}{\partial q_i} = \frac{\partial}{\partial q_i} \left(\sum_i p_i \dot{q}_i - L(q_i, \dot{q}_i, t) \right) = -\frac{\partial L}{\partial q_i} = -\frac{d}{dt}\frac{\partial L}{\partial \dot{q}_i} = -\dot{p}_i. \tag{4.15}$$

Here we have used Lagrange's equations and the definition of generalized momentum. Next, we take the partial derivative of the Hamiltonian with respect to the generalized momentum p_i:

$$\frac{\partial H}{\partial p_i} = \frac{\partial}{\partial p_i} \left(\sum_i p_i \dot{q}_i - L(q_i, \dot{q}_i, t) \right) = \dot{q}_i. \tag{4.16}$$

Equations (4.15) and (4.16) give us, once again, the canonical equations of motion:

$$\frac{\partial H}{\partial q_i} = -\dot{p}, \tag{4.17}$$

$$\frac{\partial H}{\partial p_i} = \dot{q}_i.$$

Since H is also a function of time, let us take the partial derivative of Equation (4.14) with respect to time. This yields immediately the interesting result

$$\frac{\partial H}{\partial t} = -\frac{\partial L}{\partial t}. \tag{4.18}$$

Consider the total time derivative of the Hamiltonian. It is

$$\frac{dH(q, p, t)}{dt} = \frac{\partial H}{\partial q}\frac{dq}{dt} + \frac{\partial H}{\partial p}\frac{dp}{dt} + \frac{\partial H}{\partial t}$$

$$= -\dot{p}\frac{dq}{dt} + \dot{q}\frac{dp}{dt} + \frac{\partial H}{\partial t}$$

$$= -\dot{p}\dot{q} + \dot{q}\dot{p} + \frac{\partial H}{\partial t} = \frac{\partial H}{\partial t},$$

where we used Hamilton's equations. And so

$$\frac{dH}{dt} = \frac{\partial H}{\partial t}. \tag{4.19}$$

Comparing Equations (4.18) and (4.19) we conclude that if the Lagrangian is not an explicit function of time, the Hamiltonian is constant. Furthermore, if the transformation equations do not depend explicitly on time and if the potential energy depends only on the coordinates, then the Hamiltonian is the total energy. In quantum mechanical problems one is often required to write down and solve the Schrödinger equation. This requires an expression for the Hamiltonian. Often people simply write $H = T + V$ for the Hamiltonian, and most of the time this is correct. However, you should be aware that there are situations in which the Hamiltonian is not the total energy of the system.

Exercise 4.6 Consider a frictionless harmonic oscillator consisting of a mass m on a spring of constant k, as in Section 1.10.2. Obtain the Hamiltonian, evaluate the canonical equations, and solve for the position as a function of time. (Answer: $\dot{p} = -kx$ and $\dot{x} = p_x/m$.)

4.4 Derivation of Hamilton's equations from Hamilton's principle

Hamilton's equations can also be derived by considering Hamilton's principle which states that the time development of a mechanical system is such that the action integral is stationary. That is,

$$\delta I = \delta \int_{t_1}^{t_2} L \, dt = 0.$$

We have shown that this leads to Lagrange's equations. We now show that this principle also leads to Hamilton's equations. Recall that the Lagrangian formulation treats the time development of a mechanical system as the motion of a point in *configuration* space. The Hamiltonian formulation treats the time development of a mechanical system as the motion of a point in *phase* space. For a system of N particles, phase space is a $6N$-dimensional space in which the axes are labelled q_i, p_i, $(i = 1, \ldots, n)$, where $n = 3N$. In the Hamiltonian formulation the momenta and the coordinates are on an equal footing as variables. (See Section 1.8.)

To obtain Hamilton's equations from Hamilton's principle, it must be formulated in terms of the ps and qs. Therefore, we use the definition of the

Hamiltonian as given by Equation (4.8) and write Hamilton's principle as

$$\delta I = \delta \int_{t_1}^{t_2} L \, dt = \delta \int_{t_1}^{t_2} \left(\sum_i p_i \dot{q}_i - H(q_i, p_i, t) \right) dt = 0. \qquad (4.20)$$

When written in this form, it is usually referred to as the "modified Hamilton's principle." Note that this is an expression of the form

$$\delta I = \delta \int_{t_1}^{t_2} f(q, \dot{q}, p, \dot{p}, t) \, dt = 0.$$

We know from the calculus of variations that this condition can be met if and only if the $2n$ Euler–Lagrange equations are satisfied. These equations have the form

$$\frac{d}{dt} \left(\frac{\partial f}{\partial \dot{q}_i} \right) - \frac{\partial f}{\partial q_i} = 0, \quad i = 1, \ldots, n,$$

and

$$\frac{d}{dt} \left(\frac{\partial f}{\partial \dot{p}_i} \right) - \frac{\partial f}{\partial p_i} = 0, \quad i = 1, \ldots, n.$$

Carrying out the indicated partial differentiations we obtain

$$\frac{\partial H}{\partial q_i} = -\dot{p},$$

$$\frac{\partial H}{\partial p_i} = \dot{q}_i.$$

That is, we have derived Hamilton's equations from the variational principle.

Exercise 4.7 Carry out the partial differentiation indicated above to obtain Hamilton's canonical equations.

4.5 Phase space and the phase fluid

The action integral expressed in terms of the Hamiltonian is

$$I = \int_{t_1}^{t_2} \left[\sum_{i=1}^{n} p_i \dot{q}_i - H(q_1, \ldots q_n, p_1 \ldots p_n, t) \right] dt.$$

When expressed in this way (as a function of qs and ps) it is called the "canonical integral." As we have just seen, setting the variation of this integral to zero yields the canonical equations $\dot{p}_i = -\partial H / \partial q_i$ and $\dot{q}_i = +\partial H / \partial p_i$.

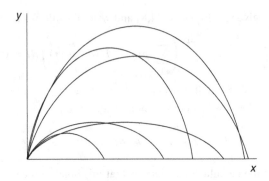

Figure 4.1   Possible trajectories for a projectile whose initial configuration space
location $(0, 0)$ is specified but whose initial velocity is not specified.

The system is now described in terms of the $2n$ variables, q_i and p_i, for $i =$
$1, \ldots, n$. The system can be represented as a point C in the $2n$-dimensional
space called phase space. (We usually refer to both the ps and qs as "coordinates.") As time goes on, point C traces out a curve in phase space.

Recall that previously we considered the system to be represented by a single
point in configuration space. (Configuration space is the n-dimensional space
made up of the qs.)

Suppose that we wanted to consider all possible paths starting from some
initial point in configuration space. Since the velocity is not specified, if we
consider all possible velocities, we would have an infinite number of curves
all starting from the same point. This is illustrated in Figure 4.1, representing
the possible trajectories of a projectile that starts from the origin but whose
velocity is not specified. Note specifically that the trajectories that originate at
a specific point in configuration space will, in general, cross trajectories that
originate at that point with a different velocity (as well as those that originate
at nearby points in configuration space).[2]

On the other hand, if we consider all possible curves from a point in *phase
space* we get a single curve because the point C now includes the momentum
of the particle as well as its position, thus giving the initial direction of motion
and the initial speed. The paths from other points will not cross the path from
our original point because the canonical equations define a unique slope at each
point in phase space. (When two trajectories cross, they have different slopes
at the point of intersection.)

---

[2] We are usually not interested in a particular solution corresponding to some special initial
conditions, but rather, a general solution that is valid for arbitrary initial conditions. The
particular solution will be a single trajectory in configuration space, but for the general
solution we obtain an infinite number of trajectories.

The totality of phase space curves originating from every point in phase space can be considered to be the *general solution* of the dynamical problem with each curve representing the motion for a particular set of initial conditions.

A different way to consider the phase space curves is to suppose that the system is made up of a great many particles. At some initial time, each particle will be at a specific point in phase space and as time goes on, each particle will trace out a curve in phase space. It is easy to picture this as analogous to the motion of water molecules in a flowing river. At some initial time, each water molecule has a specific location and a specific momentum. As time goes on the water molecules trace out streamlines that do not cross each other.

This analogy is so good that we can apply concepts of fluid dynamics to phase space and talk about the properties of the "phase fluid."

For example, the velocity of the phase fluid at every point is given by the canonical equations, $\dot{p}_i = -\partial H/\partial q_i$ and $\dot{q}_i = +\partial H/\partial p_i$. This can be interpreted as analogous to the velocity field in a real fluid. Each streamline of the moving phase fluid represents the time development of the system for given specific initial conditions and the motion of the phase fluid as a whole represents the complete solution, that is, the time development of the system for arbitrary initial conditions.

In fluid dynamics we are especially interested in *steady* fluid flows. We say that a fluid flow is steady if the velocity at a point is constant in time, that is, the velocity field does not depend on t. If the flow of our phase fluid is steady, then $\dot{p}_i$ and $\dot{q}_i$ are constants in time and the quantities $\partial H/\partial q_i$ and $\partial H/\partial p_i$ are independent of time, and H cannot contain time explicitly. That is, $\partial H/\partial t = 0$. But the total time derivative of H is

$$\frac{dH}{dt} = \sum_{i=1}^{n} \left(\frac{\partial H}{\partial q_i} \dot{q}_i + \frac{\partial H}{\partial p_i} \dot{p}_i \right) = 0,$$

by the canonical equations. So

$$\frac{dH}{dt} = 0 \Longrightarrow H = \text{constant} = E.$$

That is, energy is conserved. Furthermore, $H = E$ defines a surface in phase space and a particle will move on that surface. We conclude that, if energy is conserved, the phase fluid behaves like the steady flow of a real fluid.

As we shall see in the next chapter, there is another analogy between real fluids and the phase fluid; namely, the phase fluid will be shown to behave like an incompressible real fluid.

4.6 Cyclic coordinates and the Routhian procedure

If a coordinate does not appear in the Lagrangian it is called cyclic or ignorable. However, the Lagrangian is still a function of the corresponding generalized velocity. In other words, if one of the qs is ignorable, then the Lagrangian is a function of only $n - 1$ generalized coordinates, but it is in general a function of all n of the generalized velocities. Thus, if q_n is ignorable,

$$L = L(q_1, \ldots, q_{n-1}; \dot{q}_1, \ldots, \dot{q}_n; t).$$

However, the Hamiltonian will be a function of only $n - 1$ of the coordinates and $n - 1$ of the momenta, because the momentum conjugate to an ignorable coordinate is constant. For example, if q_n is ignorable, the corresponding momentum, p_n is a constant. Let us call it α. We can then write the Hamiltonian as

$$H = H(q_1, \ldots, q_{n-1}; p_1, \ldots, p_{n-1}; \alpha; t), \qquad (4.21)$$

where we have included the constant α in our expression even though one normally does not express the functional dependence on a constant. It is true, however, that H does depend on the value assigned to α.

The ignorable coordinate q_n does not appear in the Hamiltonian, but this does not mean that we are not interested in its time development. An ignorable coordinate is still a coordinate on the same footing as any other! Hamilton's equation for q_n is, as you might expect,

$$\dot{q}_n = \frac{\partial H}{\partial \alpha}.$$

It often happens that some of the coordinates are ignorable and others are not. For example, let us assume that a system described by the coordinates $q_1, \ldots, q_n$ is such that the first s coordinates appear in the Hamiltonian (and Lagrangian) but the coordinates $q_{s+1}, \ldots, q_n$ are cyclic. In that case a procedure for handling such problems (which was developed by Routh) can come in handy. One defines a quantity called the Routhian (which is similar in some respects to the Hamiltonian) as follows:

$$R = \sum_{i=s+1}^{n} p_i \dot{q}_i - L. \qquad (4.22)$$

Note that the summation is over the cyclic coordinates. Thus,

$$R = R(q_1, \ldots, q_s; \dot{q}_1, \ldots, \dot{q}_s; p_{s+1}, \ldots, p_n; t).$$

Taking partial derivatives of the Routhian with respect to the variables, we find

$$\frac{\partial R}{\partial q_i} = -\frac{\partial L}{\partial q_i}, \quad i = 1, \ldots, s, \tag{4.23}$$

$$\frac{\partial R}{\partial \dot{q}_i} = -\frac{\partial L}{\partial \dot{q}_i}, \quad i = 1, \ldots, s,$$

and

$$\frac{\partial R}{\partial q_i} = -\dot{p}_i, \quad i = s + 1, \ldots, n, \tag{4.24}$$

$$\frac{\partial R}{\partial p_i} = \dot{q}_i, \quad i = s + 1, \ldots, n.$$

Equations (4.23) indicate that the first s coordinates obey Lagrange's equations but with the Routhian for the *Lagrangian*, thus,

$$\frac{d}{dt}\left(\frac{\partial R}{\partial \dot{q}_i}\right) - \frac{\partial R}{\partial q_i} = 0, \quad i = 1, \ldots, s,$$

whereas Equations (4.24) indicate that the remaining coordinates and momenta obey Hamilton's equation with the Routhian as the *Hamiltonian*. But note that for the cyclic coordinates $s + 1, \ldots, n$ the momenta are constants and we can replace the $n - s$ momenta by the constants $\alpha_1, \ldots, \alpha_{n-s}$. The Routhian then can be written

$$R = R(q_1, \ldots, q_s; \dot{q}_1, \ldots, \dot{q}_s; \alpha_1, \ldots, \alpha_{n-s}; t). \tag{4.25}$$

The αs can be determined from initial conditions, and we have effectively reduced the number of variables in the problem from n to s. Reducing the number of variables is a powerful technique for solving a problem. (When we study canonical transformations we shall encounter a method for reducing the number of variables to zero!)

The Routhian is also convenient for analyzing *steady motion*. We define steady motion as motion in which the non-cyclic variables are constant. An example of steady motion is a planet moving in a *circular* orbit. In polar coordinates the Lagrangian is

$$L = \frac{1}{2}m\dot{r}^2 + \frac{1}{2}mr^2\dot{\theta}^2 + \frac{k}{r}.$$

For a circular orbit, the non-cyclic coordinate r is constant. The coordinate θ is cyclic and it increases linearly with time. The linear increase of cyclic coordinates is a characteristic of steady motion as can be appreciated from the equations of motion.

As another example, if we write the Lagrangian for a spinning, precessing top or gyroscope, we find that the only non-cyclic variable is the polar angle θ. If the top is nutating, then θ is no longer constant and the top "nods" as it precesses. This can be called "oscillation about steady motion."

In terms of our present discussion, the non-cyclic variables are constant. For a system with a Routhian of the form of Equation (4.25), the equations of motion for the cyclic coordinates (Equations (4.24)) tell us that

$$\dot{q}_i = \dot{q}_i(q_1, \ldots, q_s; \dot{q}_1, \ldots, \dot{q}_s; \alpha_1, \ldots, \alpha_{n-s}), \quad i = s+1, \ldots, n.$$

For steady motion, $q_1, \ldots, q_s$ are constant and $\dot{q}_1, \ldots, \dot{q}_s$ are zero, so $\dot{q}_i$ $(i > s) = $ constant and hence $q_i(i > s)$ varies linearly with time.

Exercise 4.8 Write the Hamiltonian for a spherical pendulum. Identify all cyclic coordinates. Write the Routhian.

Exercise 4.9 Write the Lagrangian for a spinning top in terms of Euler angles and identify the cyclic and non-cyclic coordinates.

4.7 Symplectic notation

Symplectic notation is an elegant and powerful way of expressing Hamilton's equations using matrices. One begins by defining a column matrix η whose elements are all the qs and all the ps. Specifically, for a system with n degrees of freedom, the first n elements in η are $q_1, \ldots, q_n$, and the remaining n elements are $p_1, \ldots, p_n$. Thus η has $2n$ elements. Next we define another $2n$ column matrix whose elements are the partial derivatives of the Hamiltonian with respect to all the coordinates and all the momenta so that the elements of this column matrix are

$$\frac{\partial H}{\partial q_1}, \frac{\partial H}{\partial q_2}, \ldots, \frac{\partial H}{\partial q_n}, \frac{\partial H}{\partial p_1}, \frac{\partial H}{\partial p_2}, \ldots, \frac{\partial H}{\partial p_n}.$$

We shall denote this matrix by

$$\frac{\partial H}{\partial \eta}.$$

Finally, we define a $2n \times 2n$ matrix called $\mathbf{J}$ which has the form

$$\mathbf{J} = \begin{pmatrix} 0 & 1 \\ -1 & 0 \end{pmatrix},$$

where $\mathbf{0}$ represents the $n \times n$ zero matrix and $\mathbf{1}$ represents the $n \times n$ unit matrix. Using this notation, Hamilton's equations are

$$\dot{\boldsymbol{\eta}} = \mathbf{J} \frac{\partial H}{\partial \boldsymbol{\eta}}. \tag{4.26}$$

It might be noted that such an expression lends itself to rapid computerized calculations. Computer programs that evaluate such equations of motion are referred to as "symplectic integrators" and are widely used in celestial mechanics (for the many body problem) as well as in molecular dynamics and accelerator physics.

Exercise 4.10 Show that Equation (4.26) reproduces Hamilton's equations for the simple harmonic oscillator of Exercise 4.6.

4.8 Problems

4.1 Write the Hamiltonian for the double planar pendulum.

4.2 Consider the following Lagrangian:

$$L = A\dot{x}^2 + B\dot{y}^2 + C\dot{x}\dot{y} + \frac{D}{\sqrt{x^2 + y^2}},$$

where A, B, C, D are constants. Determine the Hamiltonian.

4.3 A plane pendulum of length l and mass m is suspended in such a way that its point of support moves uniformly in a vertical circle of radius a. Obtain Hamilton's equations of motion. What is the canonical momentum?

4.4 Obtain the Hamiltonian for a planet in orbit about the Sun in plane polar coordinates. Determine Hamilton's equations of motion.

4.5 Obtain the Hamiltonian and Hamilton's equations of motion for a spherical pendulum.

4.6 Express Hamilton's equations of motion for a spherical pendulum in symplectic notation.

4.7 A particle of mass m is constrained to move on the surface of a sphere of radius R (until it falls off). The potential energy of the particle is mgz, where z is measured vertically from the plane on which the sphere is resting. Write the Hamiltonian and Hamilton's equations of motion for the time the particle is in contact with the sphere.

4.8 Consider a canonical transformation. Write the functional determinant Δ and show that $\Delta^2 = 1$. (The functional determinant is a determinant whose elements are partial derivatives of one set of variables with respect to another set. Each row contains derivatives of only one variable and each column contains derivatives with respect to only one variable.)

4.9 A massless spring of constant k is suspended from a point of support that oscillates vertically at a frequency ω such that the position of the top of the spring is given by $z_0 = A \cos \omega t$. A particle of mass m is attached to the lower end of the spring. Determine the Hamiltonian of the system.

4.10 A particle of mass m can slide freely on a horizontal rod of mass M. One end of the rod is attached to a turntable and moves in a circular path of radius a at angular speed ω. Write the Hamiltonian for the system.

5

Canonical transformations; Poisson brackets

In this chapter we begin by considering *canonical transformations.* These are transformations that preserve the form of Hamilton's equations. This is followed by a study of Poisson brackets, an important tool for studying canonical transformations. Finally we consider infinitesimal canonical transformations and, as an example, we look at angular momentum in terms of Poisson brackets.

5.1 Integrating the equations of motion

In our study of analytical mechanics we have seen that the variational principle leads to two different sets of equations of motion. The first set consists of the Lagrange equations and the second set consists of Hamilton's canonical equations. Lagrange's equations are a set of n coupled *second*-order differential equations and Hamilton's equations are a set of $2n$ coupled *first*-order differential equations.

The ultimate goal of any dynamical theory is to obtain a general solution for the equations of motion. In Lagrangian dynamics this requires integrating the equations of motion twice. This is often quite difficult because the Lagrangian (and hence the equations of motion) depends not only on the coordinates but also on their derivatives (the velocities). There is no known general method for integrating these equations.[1] You might wonder if it is possible to transform to a new set of coordinates in which the equations of motion are simpler and easier to integrate. Indeed, this is possible in some situations. But the Lagrangian is $L = T - V$ with T a function of velocities and V a function of

---

[1] Of course we can always carry out a numerical integration, but that depends on selecting a particular set of initial conditions so we do not obtain a general solution.

coordinates. One usually finds that coordinates that simplify V make T more complicated, and vice versa. Nevertheless, the idea of transforming to a different set of coordinates has merit. For example, a problem could be simplified if we transformed to a new set of coordinates in which some or all of the coordinates are *ignorable*. If a coordinate is ignorable, we have a partial integration (a "first integral") of the equation of motion because if q_i is ignorable, $p_i = \partial L/\partial \dot{q}_i$ is constant. Unfortunately, there is no known general method for transforming to a set of variables in which the coordinates are ignorable in the Lagrangian. Finding such a set of coordinates is a matter of intuition and luck rather than a mathematical procedure.

On the other hand, in Hamiltonian mechanics the situation is much brighter. Hamiltonian mechanics treats q_i and p_i as if both were coordinates. (In fact, we shall see that there are transformations that convert positions into momenta and momenta into positions.) Thus the Hamiltonian $H = H(q, p, t)$ does not depend on derivatives. Furthermore, the equations of motion are first order with all of the time derivatives isolated on one side of the equation. But most importantly, Jacobi[2] developed a technique for carrying out a transformation from q_i, p_i to a new set of coordinates (say Q_i, P_i) for which Hamilton's equations still hold and for which the integration of the equations of motion is trivial. Consequently, the problem of obtaining the integrated equations of motion is reduced to the problem of finding a "generating function" that will yield the desired transformation.

It should be mentioned that the procedure is complicated and you may find some of the concepts confusing, but this should not obscure the fact that we now have a technique for not only obtaining the equations of motion, but for actually integrating them and determining the general solution of a dynamical problem.[3]

5.2 Canonical transformations

We have seen that the transformation equations that take us from Cartesian coordinates to generalized coordinates are of the form

$$x_i = x_i(q_1, q_2, \ldots, q_n; t), \quad i = 1, \ldots, n.$$

[2] Carl Gustav Jacob Jacobi, 1804–1851.

[3] A word on terminology. The solution of a first-order differential equation that contains as many arbitrary constants as there are independent variables is called a "complete integral." If the solution depends on an arbitrary *function*, it is called the "general integral." In mechanics, one is usually interested in obtaining the complete integral.

Since such a transformation transforms the coordinates of a point from one set of coordinates to another it is called a "point transformation." (See Section 3.8.) Point transformations can be considered to take place in configuration space. In other words, a point transformation takes us from one set of configuration space coordinates (x_i) to a new set of configuration space coordinates (q_i). A point transformation yields a new set of configuration space axes.

We have also studied the Legendre transformation that allowed us to transform from the Lagrangian (a function of $q, \dot{q}, t$) to the Hamiltonian (a function of q, p, t).

We now consider a particularly important type of coordinate transformations called *canonical transformations*. These are phase space transformations that take us from the set of coordinates (p_i, q_i) to a new set (P_i, Q_i) in such a way as to *preserve the form of Hamilton's equations*. Recall that the coordinates (p_i, q_i) describe a system with Hamiltonian $H(p_i, q_i, t)$ and are such that

$$\dot{q}_i = \frac{\partial H}{\partial p_i}, \tag{5.1}$$

and

$$\dot{p}_i = -\frac{\partial H}{\partial q_i}. \tag{5.2}$$

The "new" coordinates P_i and Q_i can also be used to describe the Hamiltonian of the system. The Hamiltonian in terms of the Ps and Qs will, in general, have a different functional form than the Hamiltonian in terms of the ps and qs. Therefore, we denote it by $K(P_i, Q_i, t)$. A *canonical transformation* is one that takes us from the p_i, q_i to a set of coordinates P_i and Q_i such that

$$\dot{Q}_i = \frac{\partial K}{\partial P_i}, \tag{5.3}$$

and

$$\dot{P}_i = -\frac{\partial K}{\partial Q_i}. \tag{5.4}$$

Observe that Hamilton's equations in terms of p_i, q_i (Equations (5.1) and (5.2)) and Hamilton's equations in terms of P_i, Q_i (Equations (5.3) and (5.4)) have exactly the same *form*.

The technique for generating a canonical transformation is based on the method used to derive Hamilton's equations. Recall that they were obtained from a variational principle, specifically from the modified Hamilton's principle (Equation (4.20)), which is

$$\delta \int_{t_1}^{t_2} \left(\sum_i p_i \dot{q}_i - H(q_i, p_i, t) \right) dt = 0. \tag{5.5}$$

Hamilton's equations would certainly be obeyed if, in terms of the new coordinates, we had

$$\delta \int_{t_1}^{t_2} \left(\sum_i P_i \dot{Q}_i - K(Q_i, P_i, t) \right) dt = 0. \tag{5.6}$$

However, this is unnecessarily restrictive. For example, if F is an arbitrary function of the coordinates and time, adding a term of the form $\frac{dF}{dt}$ to the integrand of this last expression changes nothing. This is easily appreciated as follows: consider the expression

$$\delta \int_{t_1}^{t_2} \left(\sum_i P_i \dot{Q}_i - K(Q_i, P_i, t) + \frac{dF}{dt} \right) dt. \tag{5.7}$$

Since

$$\delta \int_{t_1}^{t_2} \frac{dF}{dt} dt = \delta \left(F(t_2) - F(t_1) \right),$$

and since the variation at the endpoints is required to vanish, we see that we have just added a term equal to zero to our modified Hamilton's principle. Therefore, nothing is changed, and Hamilton's principle still holds, but the new form (Equation (5.7)) allows us to generate a new Hamiltonian $K(Q_i, P_i, t)$ as well as a set of transformation equations to the new coordinates P_i and Q_i.

We know that the variation expressed by Equation (5.5) is true. The variation expressed by Equation (5.7) will certainly be true if the two integrands are equal, that is, if

$$p\dot{q} - H = P\dot{Q} - K + \frac{dF}{dt}. \tag{5.8}$$

(We have suppressed the summations and the subindices for simplicity.)

To see how Equation (5.8) leads to the transformation equations from p, q to P, Q, it is convenient to assume that F depends on a particular combination of the new and old coordinates. Thus, let us assume, for the present, that

$$F = F_1(q, Q, t). \tag{5.9}$$

(This really means $F = F_1(q_i, Q_i, t)$, $i = 1, \ldots, n$.) Note that the functional form of F is still arbitrary. All we have done is assume that F is some function of the new coordinates Q_i and the old coordinates q_i and time.

Then Equation (5.8) can be written as

$$p\dot{q} - H = P\dot{Q} - K + \frac{\partial F_1}{\partial q}\dot{q} + \frac{\partial F_1}{\partial Q}\dot{Q} + \frac{\partial F_1}{\partial t}.$$

Rearranging,

$$\left(p - \frac{\partial F_1}{\partial q}\right)\dot{q} - \left(P + \frac{\partial F_1}{\partial Q}\right)\dot{Q} = H - K + \frac{\partial F_1}{\partial t}.$$

This equation will be satisfied if

$$p = \frac{\partial F_1}{\partial q}, \tag{5.10}$$

$$P = -\frac{\partial F_1}{\partial Q}, \tag{5.11}$$

and

$$K = H + \frac{\partial F_1}{\partial t}. \tag{5.12}$$

At this stage it may appear we have merely stated the obvious, but actually we have solved the problem we set out to solve: we now have an equation for the new Hamiltonian (Equation (5.12)). Furthermore, since F_1 is a function of q and Q, then $\frac{\partial F_1}{\partial q}$ will also be a function of q and Q. So Equation (5.10) gives us an expression of the form

$$p = p(q, Q, t).$$

This can be inverted to give

$$Q = Q(p, q, t).$$

Equation (5.11) yields

$$P = P(q, Q, t),$$

but since we know $Q = Q(p, q, t)$ we can obtain

$$P = P(p, q, t).$$

The procedure we have used guarantees that Equations (5.1) and (5.2) are obeyed, so the transformation is canonical.

Since this procedure is somewhat abstract, let us carry out such a transformation for a specific choice of the generating function $F_1(q, Q, t)$. For simplicity we shall consider a system described by just two coordinates, namely p and q. Assume that

$$F_1(q, Q, t) = qQ.$$

By Equation (5.12) we see that the new Hamiltonian K is equal to the old Hamiltonian H. By Equation (5.11) we see that the new coordinate P is

$$P = -\frac{\partial F_1}{\partial Q} = -q.$$

Finally, Equation (5.10) gives

$$p = \frac{\partial F_1}{\partial q} = Q.$$

Hence we see that this particular canonical transformation leads to the new set of coordinates

$$P = -q,$$
$$Q = p.$$

In other words, the transformation changes the momentum into the coordinate and the coordinate into the momentum! This is a very good example of the fact that the Hamiltonian formulation blurs the distinction between momenta and coordinates, and that is why I referred to both of them simply as "coordinates."

The analysis we carried out assumed the generating function F was a function of the q_i and Q_i and we called it F_1. As you might expect, useful generating functions are mixed expression involving both the old variables and the new variables. There are four types of generating function to consider, and these are:

$$F = F_1(q_i, Q_i, t), \qquad (5.13)$$
$$F = F_2(q_i, P_i, t),$$
$$F = F_3(p_i, Q_i, t),$$
$$F = F_4(p_i, P_i, t).$$

For the three last types of generating function, the relations analogous to Equations (5.10), (5.11) and (5.12) are:

$$p_i = \frac{\partial F_2}{\partial q_i}, \tag{5.14}$$

$$Q_i = \frac{\partial F_2}{\partial P_i},$$

$$K = H + \frac{\partial F_2}{\partial t},$$

and

$$q_i = -\frac{\partial F_3}{\partial p_i}, \tag{5.15}$$

$$P_i = -\frac{\partial F_3}{\partial Q_i},$$

$$K = H + \frac{\partial F_3}{\partial t},$$

and

$$q_i = -\frac{\partial F_4}{\partial p_i}, \tag{5.16}$$

$$Q_i = \frac{\partial F_4}{\partial P_i},$$

$$K = H + \frac{\partial F_4}{\partial t}.$$

An interesting example of the second type of generating function is

$$F_2 = \sum_i q_i P_i. \tag{5.17}$$

In this case, as you can easily show, the new coordinates are related to the old by

$$Q_i = q_i, \tag{5.18}$$

$$P_i = p_i.$$

Thus, this is the identity transformation.

Exercise 5.1 Prove that $F_2 = \sum_i q_i P_i$ generates the identity transformation.

Exercise 5.2 Obtain the transformation Equations (5.14). Hint: Subtract PQ from F_2 before plugging into Equation (5.8).

Exercise 5.3 Let $F = F_4(p_i, P_i, t)$. Determine the new Hamiltonian and the canonical transformation equations.

Example 5.1 *Use a Canonical transformation to solve the harmonic oscilla-tor problem.*

Solution 5.1 *The Hamiltonian for a harmonic oscillator is*

$$H = \frac{1}{2m}p^2 + \frac{k}{2}q^2.$$

Using the fact that the angular frequency for a harmonic oscillator is $\omega = \sqrt{k/m}$ we can write the Hamiltonian in the more convenient form

$$H = \frac{1}{2m}\left(p^2 + m^2\omega^2 q^2\right). \tag{5.19}$$

We can solve this problem by making a canonical transformation to a new Hamiltonian $K(P, Q)$ in which the coordinate Q is cyclic. The appropriate generating function is

$$F = F_1(q, Q) = \frac{m\omega}{2}q^2 \cot Q.$$

(I have not yet shown how to obtain the appropriate generating function, so you will have to take this on faith.) Using Equations (5.10), (5.11) and (5.12) we obtain

$$p = m\omega q \cot Q,$$

$$P = \frac{m\omega q^2}{2\sin^2 Q},$$

$$K(P, Q) = H(p, q).$$

A little bit of algebra yields

$$q^2 = \frac{2}{m\omega}P\sin^2 Q,$$

and

$$p^2 = 2m\omega P \cos^2 Q.$$

These last two equations give p and q in terms of P and Q. The equation $K = H$ means that the new Hamiltonian and the old Hamiltonian have exactly the same form with p and q expressed in terms of P and Q as given by the transformation equations above. Therefore,

$$K(P, Q) = \frac{2m\omega P \cos^2 Q}{2m} + \frac{m^2\omega^2\left(\frac{2}{m\omega}P\sin^2 Q\right)}{2m}$$

$$= \omega P(\cos^2 Q + \sin^2 Q) = \omega P.$$

Thus we have obtained a Hamiltonian that is cyclic in Q, as we intended. But if the Hamiltonian is cyclic in Q then P is constant. In this case the Hamiltonian is the total energy E, so we can write

$$P = \frac{E}{\omega}.$$

Hamilton's equation for Q is

$$\dot{Q} = \frac{\partial K}{\partial P} = \omega.$$

This can be integrated immediately to yield

$$Q = \omega t + \beta,$$

where β is a constant of integration. Transforming back to our original coordinates

$$q = \sqrt{\frac{2P}{m\omega}} \sin Q,$$

so

$$q = \sqrt{\frac{2E}{m\omega^2}} \sin(\omega t + \beta),$$

and the problem is solved.

This application for solving a simple problem has been likened to using a sledgehammer to crack a peanut.[4] However, it emphasizes that in mechanics the Hamiltonian is a theoretical rather than a practical tool. (Of course, this is not true when you are doing Quantum Mechanics where the use of the Hamiltonian is the only reasonable way to solve many problems.)

Exercise 5.4 Show that, for the harmonic oscillator, $p = \sqrt{2mE} \cos(\omega t + \beta)$.

5.3 Poisson brackets

We now introduce the Poisson brackets. The notation used may be a bit confusing at first, but you will probably quickly appreciate how useful the Poisson brackets are in the development of Hamiltonian dynamics. They also

---

[4] Herbert Goldstein, *Classical Mechanics,* Addison-Wesley Pub. Co., Reading MA, USA, 1950, page 247.

pave the way for transforming classical dynamics into quantum mechanics. On a practical level, the Poisson brackets give us an easy way to determine whether or not a transformation from one set of variables to another is canonical.

Let p and q be canonical variables, and let u and v be functions of p and q. The Poisson bracket of u and v is defined as

$$[u, v]_{p,q} \equiv \frac{\partial u}{\partial q}\frac{\partial v}{\partial p} - \frac{\partial u}{\partial p}\frac{\partial v}{\partial q}. \tag{5.20}$$

Generalizing to a system of n degrees of freedom we have

$$[u, v] = \sum_{i=1}^{n}\left(\frac{\partial u}{\partial q_i}\frac{\partial v}{\partial p_i} - \frac{\partial u}{\partial p_i}\frac{\partial v}{\partial q_i}\right).$$

Using the Einstein summation convention this is written

$$[u, v] = \frac{\partial u}{\partial q_i}\frac{\partial v}{\partial p_i} - \frac{\partial u}{\partial p_i}\frac{\partial v}{\partial q_i}. \tag{5.21}$$

From the definition of the Poisson bracket it is obvious that

$$[q_i, q_j] = [p_i, p_j] = 0, \tag{5.22}$$

and that

$$[q_i, p_j] = -[p_i, q_j] = \delta_{ij}. \tag{5.23}$$

It is interesting to note that relationships involving Poisson brackets can be transformed into quantum mechanical relationships by replacing the Poisson bracket with the quantum mechanical commutator using the simple prescription

$$[u, v] \to \frac{1}{i\hbar}(\hat{u}\hat{v} - \hat{v}\hat{u}), \tag{5.24}$$

where u, v are classical functions and $\hat{u}$, $\hat{v}$ are the corresponding quantum mechanical operators.

A Poisson bracket is invariant under a change in canonical variables. That is,

$$[u, v]_{p,q} = [u, v]_{P,Q}.$$

In other words, Poisson brackets are canonical invariants. This relationship gives us an easy way to determine whether or not a set of variables is canonical.

Rules for manipulating Poisson brackets

There are a few simple rules involving Poisson brackets that can (mostly) be proved immediately by writing out the definition of the Poisson brackets. These rules essentially define the arithmetic of Poisson brackets. They are:

$$[u, u] = 0 \tag{5.25}$$

$$[u, v] = -[v, u] \tag{5.26}$$

$$[au + bv, w] = a[u, w] + b[v, w] \tag{5.27}$$

$$[uv, w] = [u, w]v + u[v, w]. \tag{5.28}$$

A useful relationship which is somewhat harder to prove is called Jacobi's identity. It is:

$$[u, [v, w]] + [v, [w, u]] + [w, [u, v]] = 0. \tag{5.29}$$

Exercise 5.5 Prove Equations (5.22) and (5.23).

Exercise 5.6 Prove that Poisson brackets are invariant under canonical transformations.

Exercise 5.7 Prove Equations (5.25) and (5.26).

5.4 The equations of motion in terms of Poisson brackets

We have expressed the equations of motion for a mechanical system in a variety of ways, including Newton's second law, the Lagrange equations, and Hamilton's equations. The equations of motion can also be expressed in terms of the Poisson brackets, as we now demonstrate.

Recall that Hamilton's equations are a set of $2n$ first-order equations for $\dot{q}_i$ and $\dot{p}_i$, that is, for the first derivative of the variables. (This differs from the Lagrange equations or Newton's laws which are a set of n second-order equations for $\ddot{q}_i$.)

If u is a function of q_i, p_i, t, we can write

$$\frac{du}{dt} = \frac{\partial u}{\partial q_i}\dot{q}_i + \frac{\partial u}{\partial p_i}\dot{p}_i + \frac{\partial u}{\partial t}.$$

But $\dot{q}_i$ and $\dot{p}_i$ can be expressed using Hamilton's equations so we obtain

$$\frac{du}{dt} = \frac{\partial u}{\partial q_i}\frac{\partial H}{\partial p_i} - \frac{\partial u}{\partial p_i}\frac{\partial H}{\partial q_i} + \frac{\partial u}{\partial t},$$

$$= [u, H] + \frac{\partial u}{\partial t}. \tag{5.30}$$

If $u =$ constant, $\frac{du}{dt} = 0$ and $[u, H] = -\frac{\partial u}{\partial t}$. Therefore, if u does not depend explicitly on t, then $[u, H] = 0$.[5]

Now suppose that $u = q$. We have

$$\dot{q} = [q, H]. \tag{5.31}$$

Similarly,

$$\dot{p} = [p, H]. \tag{5.32}$$

Thus we have written Hamilton's equations of motion in terms of Poisson brackets. Furthermore, if we let $u = H$ we have

$$\frac{dH}{dt} = [H, H] + \frac{\partial H}{\partial t},$$

or

$$\frac{dH}{dt} = \frac{\partial H}{\partial t}, \tag{5.33}$$

as we have already seen (Equation (4.19)).

In terms of symplectic notation where the column matrix η has components $q_1, \ldots, p_n$, we can combine Equations (5.31) and (5.32) into one, thus

$$\dot{\eta} = [\eta, H], \tag{5.34}$$

which is a prescription for generating the values of $q_1, \ldots, p_n$ at some later time if their values at time t are known, because

$$\eta(t + dt) = \eta(t) + \dot{\eta}dt.$$

5.4.1 Infinitesimal canonical transformations

In an infinitesimal canonical transformation the new coordinates differ only infinitesimally from the old coordinates. Therefore, the transformation equations from p_i, q_i to P_i, Q_i will have the form

---

[5] In quantum mechanical terminology we would say that $\hat{u}$ commutes with $\hat{H}$. See Equation (5.24). You probably recall from your quantum mechanics course that conserved quantities commute with the Hamiltonian.

$$Q_i = q_i + \delta q_i,$$

$$P_i = p_i + \delta p_i.$$

(A word on the notation. Here δq_i and δp_i are infinitesimal changes in the q_i and p_i and are *not* virtual displacements.)

For the identity transformation we wrote

$$p_i = \frac{\partial F_2}{\partial q_i},$$

$$Q_i = \frac{\partial F_2}{\partial P_i},$$

with $F_2 = \sum q_i P_i$. (See Equations (5.14).) Therefore, the infinitesimal canonical transformation which differs infinitesimally from the identity transformation has a "generating function" given by

$$F_2(q, P, t) = \sum q_i P_i + \epsilon G(q, P, t),$$

where ϵ is an infinitesimal quantity and G is an arbitrary function. Then, by the rules for F_2 transformations (Equations (5.14)) we see that

$$p_j = \frac{\partial F_2}{\partial q_j} = P_j + \epsilon \frac{\partial G}{\partial q_j},$$

$$Q_j = \frac{\partial F_2}{\partial P_j} = q_j + \epsilon \frac{\partial G}{\partial P_j}.$$

Now,

$$\delta q_j = Q_j - q_j = \epsilon \frac{\partial G}{\partial P_j},$$

and to first order in ϵ we can write

$$\delta q_j = \epsilon \frac{\partial G}{\partial p_j}. \tag{5.35}$$

Similarly, $\delta p_j = P_j - p_j$ so

$$\delta p_j = -\epsilon \frac{\partial G}{\partial q_j}. \tag{5.36}$$

In symplectic notation Equations (5.35) and (5.36) are expressed as

$$\delta \eta = \epsilon [\eta, G]. \tag{5.37}$$

Since F_2 is guaranteed to generate a canonical transformation, we appreciate that (5.37) is indeed an infinitesimal canonical transformation.

In general

$$\zeta = \eta + \delta\eta, \tag{5.38}$$

is a transformation from $q_1, \ldots, p_n$ to a nearby point in phase space $q_1 + \delta q_1, \ldots, p_n + \delta p_n$. But if the quantity G is taken to be the Hamiltonian H, the transformation is a displacement in *time* from $q_1(t), \ldots, p_n(t)$ to $q_1(t + dt), \ldots, p_n(t + dt)$. (This canonical transformation in time is sometimes called a "contact" transformation.) Thus, the Hamiltonian is the generator of the motion of the system in time; it generates the canonical variables at a later time.

We now consider the change in a function $u = u(q, p)$ under a canonical transformation. Note that there are two ways to interpret a canonical transformation. On the one hand, it changes the description of a system from an old set of canonical variables q, p to a new set of canonical variables Q, P. Under such a transformation a function u may have a different *form* but it will still have the same *value*. Thus, if $q_1(t), \ldots, p_n(t)$ are denoted by A and $Q_1(t), \ldots, P_n(t)$ are denoted by A', the transformed function $u(A')$ has the same value as the original function $u(A)$:

$$u(A') = u(A).$$

(The value of the total angular momentum of a rotating body is the same regardless of the set of coordinates used to evaluate it.) Of course, $u(A')$ will in general have a different mathematical form than $u(A)$. (In Cartesian coordinates the angular momentum of a mass point moving in a circle is $m(x\dot{y} - y\dot{x})$ and in polar coordinates it is $mr^2\dot{\theta}$.)

On the other hand, if we consider an infinitesimal contact transformation that generates a translation in *time*, the interpretation is quite different. Now the transformation takes us from A, representing the values $q_1(t), \ldots, p_n(t)$, to B, representing $q_1(t + dt), \ldots, p_n(t + dt)$, in the same phase space (with the same set of phase space axes) but at a later time. Since we are in the same phase space and using the same set of coordinates, the *form* of u is conserved but its *value* changes. Let us represent such a change in the function u by

$$\partial u = u(B) - u(A).$$

But since $u = u(q, p)$ we write

$$
\begin{aligned}
\partial u &= \frac{\partial u}{\partial q_i} \delta q_i + \frac{\partial u}{\partial p_i} \delta p_i \\
&= \frac{\partial u}{\partial q_i} \epsilon \frac{\partial G}{\partial pi} + \frac{\partial u}{\partial p_i} \left(-\epsilon \frac{\partial G}{\partial q_i} \right) \\
&= \epsilon \left[\frac{\partial u}{\partial q_i} \frac{\partial G}{\partial p_i} - \frac{\partial u}{\partial p_i} \frac{\partial G}{\partial q_i} \right],
\end{aligned}
$$

or

$$
\partial u = \epsilon [u, G]. \tag{5.39}
$$

Now let us consider a change in the Hamiltonian under a contact transformation. The Hamiltonian differs from an ordinary function u in the following way. Under a canonical transformation $u(A) \rightarrow u(A')$ the value of u is constant, but for the Hamiltonian, a canonical transformation from A to A' can result in a function with a different value. In fact, the transformed Hamiltonian will generally be a completely different function. This is because the Hamiltonian is not a function having a specific value, but rather a function which defines the canonical equations of motion. Therefore, we must write

$$
H(A) \rightarrow K(A').
$$

As we know,

$$
K = H + \frac{\partial F}{\partial t}.
$$

But for an infinitesimal canonical transformation, we want K to differ from H by only an infinitesimal amount, so F is approximately equal to the identity transformation. As before, we write

$$
F_2 = \sum q_i P_i + \epsilon G(q, P, t).
$$

Then,

$$
K(A') = H(A) + \frac{\partial}{\partial t} \left(\sum q_i P_i + \epsilon G(q, P, t) \right) = H(A) + \epsilon \frac{\partial G}{\partial t},
$$

and

$$
\partial H = H(B) - K(A') = H(B) - H(A) - \epsilon \frac{\partial G}{\partial t}.
$$

But

$$
H(B) - H(A) = \epsilon [H, G],
$$

so

$$\partial H = \epsilon[H, G] - \epsilon \frac{\partial G}{\partial t}.$$

Now by Equation (5.30),

$$\frac{dG}{dt} = [G, H] + \frac{\partial G}{\partial t}.$$

Therefore

$$\epsilon \frac{\partial G}{\partial t} = \epsilon \frac{dG}{dt} - \epsilon[G, H],$$

and

$$\partial H = \epsilon[H, G] - \epsilon \frac{\partial G}{\partial t} + \epsilon[G, H] = -\epsilon \frac{\partial G}{\partial t}. \tag{5.40}$$

If G is a constant of the motion, $\frac{dG}{dt} = 0$ and $\partial H = 0$. That is, if the generating function is a constant of the motion, the Hamiltonian is invariant. This is closely related to the relationship between symmetries and constants of the motion. Recall that under a translation involving an ignorable coordinate, the Hamiltonian is invariant and the conjugate generalized momentum is constant. But if the Hamiltonian is invariant, then G, the generator of the infinitesimal canonical transformation, must be a constant of the motion (because then $\frac{dG}{dt} = 0$). That is, the symmetry (the cyclic nature of one of the coordinates) implies a constant of the motion (G).

5.4.2 Canonical invariants

So far we have considered two quantities that are canonical invariants, that is, quantities that do not change when the variables undergo a canonical transformation from the p, q to the P, Q. The first set of such canonical invariants is, of course, Hamilton's equations. The second is the Poisson bracket. Any equation expressed in terms of Poisson brackets will be invariant under a canonical transformation of variables. One can actually develop a system of mechanics based on Poisson brackets in which all equations are the same for any set of canonical variables.

A third quantity that can be shown to be a canonical invariant is the volume element in phase space. In other words, if the volume element in terms of the variables q_i, p_i is

$$d\eta = dq_1 dq_2 \cdots dq_n \, dp_1 dp_2 \cdots dp_n,$$

and the volume element in terms of the canonical variables Q_i, P_i is

$$d\zeta = dQ_1 dQ_2 \cdots dQ_n \, dP_1 dP_2 \cdots dP_n,$$

then

$$d\eta = d\zeta. \tag{5.41}$$

Let us consider the proof of this assertion. The proof is based on the fact that the transformation of the infinitesimal volume element expressed in one set of variables $(x_1, x_2, \ldots, x_n)$ to another set $(q_1, q_2, \ldots, q_m)$ is given by

$$dx_1 dx_2 \cdots dx_n = D dq_1 dq_2 \cdots dq_m,$$

where D is the Jacobian determinant defined by

$$D = \frac{\partial(q_1, \ldots, q_m)}{\partial(x_1, \ldots, x_n)} = \begin{vmatrix} \frac{\partial q_1}{\partial x_1} & \frac{\partial q_1}{\partial x_2} & \cdots & \frac{\partial q_1}{\partial x_n} \\ & & \vdots & \\ \frac{\partial q_m}{\partial x_1} & \frac{\partial q_m}{\partial x_2} & \cdots & \frac{\partial q_m}{\partial x_n} \end{vmatrix}.$$

The volume of some region of phase space expressed in terms of the two sets of canonical variables $(q_1, \ldots, p_n)$ and $(Q_1, \ldots, P_n)$ is

$$\int \int \cdots \int dQ_1 dQ_2 \cdots dP_n = \int \int \cdots \int D dq_1 dq_2 \cdots dp_n.$$

The volume is invariant if and only if $D = 1$. Thus the proof of the invariance of the volume under a canonical transformation is reduced to proving that the Jacobian determinant is equal to unity.

Note that

$$D = \frac{\partial(Q_1, \ldots, P_n)}{\partial(q_1, \ldots, p_n)} = \begin{vmatrix} \frac{\partial Q_1}{\partial q_1} & \frac{\partial Q_1}{\partial q_2} & \cdots & \frac{\partial Q_1}{\partial p_n} \\ & & \vdots & \\ \frac{\partial P_n}{\partial q_1} & \frac{\partial P_n}{\partial q_2} & \cdots & \frac{\partial P_n}{\partial p_n} \end{vmatrix}.$$

Now the Jacobian can be treated somewhat like a fraction and the value of D is not changed if we divide top and bottom by $\partial(q_1, \ldots, q_n, P_1, \ldots, P_n)$, thus

$$D = \frac{\partial(Q_1, \ldots, P_n)}{\partial(q_1, \ldots, p_n)} = \frac{\frac{\partial(Q_1, \ldots, P_n)}{\partial(q_1, \ldots, q_n, P_1, \ldots, P_n)}}{\frac{\partial(q_1, \ldots, p_n)}{\partial(q_1, \ldots, q_n, P_1, \ldots, P_n)}}.$$

Another property of Jacobians (which follows from the properties of determinants) is that if the same quantities appear in the numerator and

denominator, they can be dropped, yielding a Jacobian of lower order.[6]
Therefore we can write

$$D = \frac{\partial(Q_1, \ldots, Q_n)/\partial(q_1, \ldots, q_n)}{\partial(p_1, \ldots, p_n)/\partial(P_1, \ldots, P_n)} = \frac{|n|}{|d|},$$

where we dropped the Ps in the numerator and the qs in the denominator.
(The quantity $|n|$ represents the determinant of the numerator and $|d|$ represents the determinant of the denominator.) The ikth element of the determinant in the numerator is

$$n_{ik} = \frac{\partial Q_i}{\partial q_k},$$

and the kith element of the denominator is

$$d_{ki} = \frac{\partial p_k}{\partial P_i}.$$

If the generating function of the canonical transformation has the form
$F = F_2(q_i, P_i, t)$, then by Equations (5.14)

$$p_i = \frac{\partial F}{\partial q_i},$$

$$Q_i = \frac{\partial F}{\partial P_i},$$

and consequently the ikth element of the numerator is

$$n_{ik} = \frac{\partial}{\partial q_k} \frac{\partial F}{\partial P_i} = \frac{\partial^2 F}{\partial q_k \partial P_i},$$

and the kith element of the denominator is

$$d_{ki} = \frac{\partial}{\partial P_i} \frac{\partial F}{\partial q_k} = \frac{\partial^2 F}{\partial q_k \partial P_i},$$

so the two determinants differ only in the fact that the rows and columns are interchanged. This does not affect the value of a determinant, so we conclude that $D = 1$ and the assertion is proved.

Exercise 5.8 Show that for the transformation from Cartesian to polar coordinates that $D = r$.

---

[6] See, for example, Section 6.4.4 of *Mathematical Methods for Physics and Engineering* by K. F. Riley, M. P. Hobson and S. J. Bence, 2nd Edn, Cambridge University Press, 2002.

Exercise 5.9 Consider a canonical transformation from (q, p) to (Q, P). (Here $n = 1$.) (a) Show that

$$\frac{\partial(Q, P)}{\partial(q, p)} = \frac{\partial(Q, P)/\partial(q, P)}{\partial(q, p)/\partial(q, P)}.$$

(b) Show that

$$\frac{\partial(Q, P)}{\partial(q, P)} = \frac{\partial Q}{\partial q}.$$

5.4.3 Liouville's theorem

Recall the analogy between the phase fluid and real fluids and consider that some fluids are incompressible. (Water is a reasonably good example of an incompressible fluid.) From fluid dynamics it is known that if a fluid is incompressible, then the velocity field $\mathbf{v} = \mathbf{v}(x, y, z)$ has zero divergence: $\nabla \cdot \mathbf{v} = 0$. The phase fluid behaves like a $2n$-dimensional incompressible fluid. The "velocity" components for this fluid are $\dot{q}_i$ and $\dot{p}_i$. Therefore, the condition for incompressibility is

$$\sum_{i=1}^{n} \left(\frac{\partial \dot{q}_i}{\partial q_i} + \frac{\partial \dot{p}_i}{\partial p_i} \right) = 0. \tag{5.42}$$

Applying the divergence theorem to the velocity of an incompressible fluid we find that the flux of the fluid across any closed surface is zero. Similarly, the flux of the phase fluid over any closed surface in phase space is zero. This fact, which was discovered by Liouville, is called "Liouville's theorem." Since the development of a system in time can be treated as a canonical transformation, this is just another way of stating that the volume of a region of the phase fluid is constant in time, and we have just proved that this is a property of canonical transformations.

Liouville's theorem is particularly useful in statistical mechanics. For example, in considering a gas the system will probably contain about 10^{26} molecules. It is clearly out of the question to follow the motion of each individual particle. In statistical mechanics this type of problem is dealt with by imagining an *ensemble* of such systems and evaluating ensemble averages of the parameters of interest. For example, the ensemble could consist of a large number of identical systems of gas molecules differing only in their initial conditions. In phase space, each member of the ensemble is represented by a single point and the whole ensemble is represented by a large number of points. Liouville's theorem states that the density of these points remains constant in time. Note that if the phase fluid is incompressible, it has a constant density. By the constant density of the phase fluid we mean that the number of points in a given volume does not change. Each point represents a system. As time goes on, the points in a volume move along their phase paths and the surface enclosing them will be distorted, but it will have the same volume as before. Now the volume of a region of phase space is given by

$$\tau = \int dq_1 \cdots dq_n dp_1 \cdots dp_n.$$

Liouville's theorem gives us an additional constant of the motion, namely

$$\tau = \text{constant.}$$

Another point of view is to consider an infinitesimal region of phase space containing a number of systems. The density is just the number of systems divided by the volume. At a later time, these systems will have moved to a different region of phase space. The region enclosing the points will have changed shape, but will have the same volume as before because the volume element in phase space is a canonical invariant. But if the number of systems is constant and the volume is constant, then the density is constant and the theorem is proved.

It is interesting to note that if the density is denoted by ρ then in general,

$$\frac{d\rho}{dt} = [\rho, H] + \frac{\partial \rho}{\partial t}.$$

But since $\frac{d\rho}{dt} = 0$ we have

$$\frac{\partial \rho}{\partial t} = -[\rho, H].$$

Exercise 5.10 Show that if $\nabla \cdot \mathbf{v} = 0$ then there is no net flow of material across the boundary of a given volume. (Hint: Use the divergence theorem.)

Exercise 5.11 Show that Equation (5.42) is true in general.

5.4.4 Angular momentum

Another application of infinitesimal canonical transformations involves the angular momentum of a system. If a system is symmetric with respect to a rotation, then the rotation angle (say θ) is cyclic and the conjugate generalized momentum (the angular momentum **L**) will be constant.[7] Conversely, if the generating function is taken to be the angular momentum, the transformation is a rigid rotation of the system. This can be shown as follows. Let

$$G = L_z = \sum (\mathbf{r}_i \times \mathbf{p}_i)_z = \sum (x_i p_{yi} - y_i p_{xi}).$$

(We selected the z component of angular momentum for simplicity; this does not imply any lack of generality.) If we carry out an infinitesimal rotation $d\theta$ about the z axis, it is easy to show that

$$\delta x_i = -y_i d\theta$$
$$\delta y_i = x_i d\theta$$
$$\delta z_i = 0,$$

and

$$\delta p_{xi} = -p_{yi} d\theta$$
$$\delta p_{yi} = p_{xi} d\theta$$
$$\delta p_{zi} = 0.$$

Now we have seen that $\delta q_j = \epsilon \frac{\partial G}{\partial p_j}$ (Equation (5.35)) and that $\delta p_j = -\epsilon \frac{\partial G}{\partial q_j}$ (Equation (5.36)), so if we use $d\theta$ for the infinitesimal parameter ϵ,

$$\delta x_i = \epsilon \frac{\partial G}{\partial p_{xi}} = \epsilon \frac{\partial}{\partial p_{xi}} \left(\sum_j x_j p_{yj} - y_j p_{xj} \right) = \epsilon(-y_i) = -y_i d\theta,$$

as expected. The other relations follow similarly.

5.5 Angular momentum in Poisson brackets

We have seen that the infinitesimal canonical transformation generates a change in a function u given by

$$\partial u = \epsilon[u, G].$$

In the "active" interpretation we think of the ICT as a transformation that generates the (new) value of the function due to some change in the system parameters. In particular, if G is the Hamiltonian, then we let $\epsilon = dt$ and ∂u is the change in the function u as the parameter *time* goes from t to $t + dt$. If G is some other function, then ∂u represents a different sort of change in u.

---

[7] Do not confuse the vector angular momentum **L** with the Lagrangian.

If u is a particular component of a vector $\mathbf{F}$, say F_i, then the change in F_i due to an infinitesimal canonical transformation is

$$\partial F_i = \epsilon[F_i, G].$$

If G is a component of the angular momentum ($G = \hat{\mathbf{n}} \cdot \mathbf{L}$) and ϵ is taken to be $d\theta$, then the ICT gives the change in F_i due to a rotation of the system through the angle $d\theta$ about $\hat{\mathbf{n}}$:

$$\partial F_i = d\theta[F_i, \hat{\mathbf{n}} \cdot \mathbf{L}].$$

Note that this ICT gives the change in the component F_i along some direction fixed in space, that is, along a fixed axis. For example, F_i might be $\mathbf{F} \cdot \hat{k}$ where the Cartesian unit vectors $\hat{\imath}, \hat{\jmath}, \hat{k}$ are assumed to be fixed in space and not affected by the rotation. Clearly, the components of some vectors are affected by the rotation and others are not. Axes fixed in the body rotate with it; axes fixed in space do not. An external vector, such as the gravitational force, is obviously not affected by a rotation of the body. On the other hand, a component of the angular momentum of the system would, in general, be affected by such a rotation. A "system vector" is a vector whose components depend on the orientation of the body and which is affected by a rotation of the system.

The change in a system vector due to an infinitesimal rotation $d\theta$ about some axis $\hat{\mathbf{n}}$ is, according to vector analysis,

$$d\mathbf{F} = \hat{\mathbf{n}}d\theta \times \mathbf{F},$$

whereas the Poisson bracket formulation yields

$$\partial \mathbf{F} = d\theta[\mathbf{F}, \hat{\mathbf{n}} \cdot \mathbf{L}].$$

The two formulations must be equivalent, so

$$[\mathbf{F}, \hat{\mathbf{n}}\cdot\mathbf{L}] = \hat{\mathbf{n}} \times \mathbf{F}. \tag{5.43}$$

This equation is particularly interesting and useful because all reference to the rotation has been deleted and is valid for all system vectors. It gives the relationship between the Poisson bracket of the system vector with a component of the angular momentum $[\mathbf{F}, \hat{\mathbf{n}}\cdot\mathbf{L}]$ and the cross product of the unit vector in the direction of the angular momentum component and the system vector $\hat{\mathbf{n}}\times\mathbf{F}$. Sometimes this last relationship is more easily used in terms of components in which case it reads

$$[F_i, L_j] = \epsilon_{ijk} F_k, \tag{5.44}$$

where ϵ_{ijk} is the Levi–Civita density.[8]

Another useful relationship involves the Poisson bracket of the dot product of two system vectors (say **F** and **V**) and a component of **L**. Using (5.43):

$$[\mathbf{F} \cdot \mathbf{V}, \hat{\mathbf{n}} \cdot \mathbf{L}] = \mathbf{F} \cdot [\mathbf{V}, \hat{\mathbf{n}} \cdot \mathbf{L}] + \mathbf{V} \cdot [\mathbf{F}, \hat{\mathbf{n}} \cdot \mathbf{L}] \tag{5.45}$$

$$= \mathbf{F} \cdot (\hat{\mathbf{n}} \times \mathbf{V}) + \mathbf{V} \cdot (\hat{\mathbf{n}} \times \mathbf{F})$$

$$= (\mathbf{F} \times \hat{\mathbf{n}}) \cdot \mathbf{V} + \mathbf{V} \cdot (\hat{\mathbf{n}} \times \mathbf{F}) = 0,$$

where we exchanged the dot and cross on the first term. Now if **F** and **V** are both taken to be **L**, Equation (5.44) gives the following relation between the components of the angular momentum

$$[L_i, L_j] = \epsilon_{ijk} L_k, \tag{5.46}$$

and Equation (5.45) gives

$$[\mathbf{L}^2, \hat{\mathbf{n}} \cdot \mathbf{L}] = 0. \tag{5.47}$$

It can easily be shown that the Poisson bracket of any two constants of the motion is also a constant of the motion. (This is called Poisson's theorem.) Therefore, if L_x and L_y are constants of the motion, then by (5.46), so too is L_z. Recall that according to Equation (5.22) the Poisson bracket of any two canonical momentum components is zero. But $[L_i, L_j] = \epsilon_{ijk} L_k$, so L_i and L_j cannot be canonical variables. More generally, no two components of **L** can simultaneously be canonical variables. One the other hand, the relation $[\mathbf{L}^2, \hat{\mathbf{n}} \cdot \mathbf{L}] = 0$ shows that the magnitude of L (or L^2) and a component of **L** can simultaneously be canonical variables.[9]

Furthermore if **p** is the linear momentum of the system and if p_z is constant, then so too are p_x and p_y. This is easily proved because by Equation (5.44)

$$[p_z, L_x] = p_y,$$

and, according to Poisson's theorem, if p_z and L_x are constants, then p_y is also a constant. Similarly,

$$[p_z, L_y] = -p_x,$$

---

[8] The Levi-Civita density ϵ_{ijk} is defined to be zero if any two indices are equal; it is $+1$ if ijk are even permutations of 123, and it is -1 if ijk are odd permutations of 123.

[9] This statement carries over into quantum mechanics where we find that no two components of **L** can simultaneously have eigenvalues, but, for example, L_z and L^2 can be quantized together.

and so p_x is also constant. Thus, both L and $\mathbf{p}$ are conserved if L_x, L_y and p_z are constant.

All of these relationships are probably familiar to you from your study of quantum mechanics.

5.6 Problems

5.1 (a) Prove that the transformation

$$Q = p/\tan q,$$
$$P = \log(\sin q/p),$$

is canonical. (b) Find the generating function $F_1(q, Q)$ for this transformation.

5.2 Assume $H(q_1, q_2, p_1, p_2) = q_1 p_1 - q_2 p_2 - aq_1^2 + bq_2^2$, where a and b are constants. Show that the time derivative of $q_1 q_2$ is zero. (Use Poisson brackets.)

5.3 Under what condition is the following transformation not canonical?

$$Q = q + ip,$$
$$P = Q^*.$$

5.4 Consider the generating function

$$F_2(q, P) = (q + P)^2.$$

Obtain the transformation equations.

5.5 The Hamiltonian for a falling body is

$$H = \frac{p^2}{2m} + mgz,$$

where g is the acceleration of gravity. Determine the generating function $F_4(p, P)$ if $K = P$.

5.6 Does $F_1(q, Q) = q^2 + Q^4$ generate a canonical transformation?

5.7 Suppose f and g are constants of the motion. Prove that their Poisson bracket is also an integral of the motion.

5.8 Prove Jacobi's identity (Equation (5.29)). Show that the commutator $[\mathbf{A}, \mathbf{B}] = \mathbf{AB} - \mathbf{BA}$ also satisfies a relation similar to the Jacobi identity. That is, show that $[\mathbf{A}, \mathbf{BC}] = [\mathbf{A}, \mathbf{B}]\mathbf{C} + \mathbf{B}[\mathbf{A}, \mathbf{C}]$.

5.9 Consider a generating function of the form $F_3(p_i, Q_i, t)$. (a) Derive expressions for q_i, P_i and K, as given by Equations (5.15). (b) Determine q, P, and K, if $F_3 = pQ$.

5.10 Assume q and p are canonical. Show that if Q and P are canonical, then the following conditions hold

$$\frac{\partial Q}{\partial q} = \frac{\partial p}{\partial P}, \quad \frac{\partial Q}{\partial p} = -\frac{\partial q}{\partial P}, \quad \frac{\partial P}{\partial q} = -\frac{\partial p}{\partial Q}, \quad \frac{\partial P}{\partial p} = \frac{\partial q}{\partial Q}.$$

Prove directly and by using Poisson brackets.

6

Hamilton–Jacobi theory

In the previous chapter it was mentioned that there is no general technique for solving the n coupled second-order Lagrange equations of motion, but that Jacobi had derived a general method for solving the $2n$ coupled canonical equations of motion, allowing one to determine all the position and momentum variables in terms of their initial values and the time.

There are two slightly different ways to solve Hamilton's canonical equations. One is more general, whereas the other is a bit simpler, but is only valid for systems in which energy is conserved. We will go through the procedure for the more general method, then solve the harmonic oscillator problem by using the second method.

Both methods involve solving a partial differential equation for the quantity S that is called "Hamilton's principal function." The problem of solving the entire system of equations of motion is reduced to solving a single partial differential equation for the function S. This partial differential equation is called the "Hamilton–Jacobi equation." Reducing the dynamical problem to solving just one equation is quite satisfying from a theoretical point of view, but it is not of much help from a practical point of view because the partial differential equation for S is often very difficult to solve. Problems that can be solved by obtaining the solution for S can usually be solved more easily by other means.

Nevertheless, the solution of the Hamilton–Jacobi equation is important for a variety of reasons. For one thing, the analysis allows one to draw an "optical-mechanical analogy" between optical rays and mechanical phase space paths. That is, we can treat waves from a mechanical point of view. This led, of course, to "wave mechanics" and quantum theory.

6.1 The Hamilton–Jacobi equation

A dynamical problem is solved when we determine relationships for the generalized coordinates and momenta in terms of the initial conditions and the time.
The initial conditions can be expressed as q_0 and p_0, by which we mean the
constant values of the coordinates (q_i) and the momenta (p_i) at time $t = 0$.
The solution of the mechanical problem is the set of relationships

$$q = q(q_0, p_0, t), \tag{6.1}$$
$$p = p(q_0, p_0, t).$$

These equations can be considered to be a set of *transformation equations*
between the canonical variables q, p and a new set q_0, p_0. Finding a solution
to the mechanical problem is equivalent to finding the transformation from q, p
to q_0, p_0. But q_0, p_0 are constants; therefore we need to find a transformation
from the canonical variables (q, p) to a set of variables which are constants.
We now show how this can be done.

A canonical transformation takes us from the Hamiltonian $H = H(p, q, t)$
to the transformed Hamiltonian $K = K(P, Q, t)$, where

$$\dot{Q}_i = \frac{\partial K}{\partial P_i},$$
$$\dot{P}_i = -\frac{\partial K}{\partial Q_i}$$

and

$$K = H + \frac{\partial F}{\partial t}.$$

If the new variables are constants we have $\dot{Q}_i = 0$ and $\dot{P}_i = 0$. (We want
the "new" variables to be the constant initial values of the coordinates and
momenta.) The requirement that the new variables be constant is guaranteed
quite simply by requiring that K be a constant. In fact, if we set $K = 0$, then
obviously $\dot{Q}_i = 0$ and $\dot{P}_i = 0$. Setting $K = 0$ yields

$$H + \frac{\partial F}{\partial t} = 0. \tag{6.2}$$

This is a differential equation that can be solved for F. (Recall that F is the
generating function for the canonical transformation.)

We shall chose F to have the form $F = F_2(q, P, t)$. This is a reasonable
choice because this generating function takes us from p to P and from q to Q.
You will also recall that this is the basis for the infinitesimal contact transformations that took us from q and p at one time to q and p at some other time.

We previously denoted this generating function by F_2, but now, for historical reasons, we shall denote it by S. We do not (yet) know the functional form of S, but we do know that

$$S = S(q, P, t).$$

In Hamilton–Jacobi theory, the function S is called Hamilton's principal function.

By Equations (5.14)

$$p_i = \frac{\partial S}{\partial q_i}, \tag{6.3}$$

$$Q_i = \frac{\partial S}{\partial P_i},$$

$$K = H + \frac{\partial S}{\partial t}.$$

If we can determine S, then we will be able to generate the transformation equations (6.1) and the problem is solved.

Writing (6.2) in terms of S yields the following relationship

$$H(q_1, \ldots, q_n; p_1, \ldots, p_n; t) + \frac{\partial S}{\partial t} = 0.$$

But $p_i = \frac{\partial S}{\partial q_i}$, so we can write

$$H(q_1, \ldots, q_n; \frac{\partial S}{\partial q_1}, \ldots, \frac{\partial S}{\partial q_n}; t) + \frac{\partial S}{\partial t} = 0. \tag{6.4}$$

This is a partial differential equation for S. It is called the *Hamilton–Jacobi equation*. Given the functional form of H it can be solved for S.

Recall that the new variables P and Q are the constant initial conditions. To emphasize this, one usually writes α_i for P_i and β_i for Q_i. Then,

$$S = S(q_i, \alpha_i, t).$$

By Equations (6.3)

$$p_i = \frac{\partial S}{\partial q_i} = \frac{\partial S(q_i, \alpha_i, t)}{\partial q_i}, \tag{6.5}$$

and

$$\beta_i = Q_i = \frac{\partial S}{\partial \alpha_i} = \frac{\partial S(q_i, \alpha_i, t)}{\partial \alpha_i}. \tag{6.6}$$

If Equation (6.6) can be inverted we obtain

$$q_i = q_i(\alpha_i, \beta_i, t). \tag{6.7}$$

Plugging (6.7) into (6.5) gives

$$p_i = \frac{\partial}{\partial q_i} S(q_i(\alpha_i, \beta_i, t), \alpha_i, t),$$

or

$$p_i = p_i(\alpha_i, \beta_i, t). \tag{6.8}$$

Equations (6.7) and (6.8) are the desired solution to the dynamical problem.

Exercise 6.1 Write the Hamilton–Jacobi equation for a free particle.

Exercise 6.2 Determine Hamilton's principal function for a free particle.

6.2 The harmonic oscillator – an example

As an example of the use of Hamilton–Jacobi theory, let us apply it to the harmonic oscillator problem. Recall that the Hamiltonian for the harmonic oscillator (see Equation 5.19) can be written as

$$H = \frac{1}{2m} \left(p^2 + m^2 \omega^2 q^2 \right),$$

where $\omega = \sqrt{k/m}$. Since H does not contain the time t explicitly, H is a constant and in this case we know it is equal to the total energy ($H = E$).

The Hamilton–Jacobi equation (6.4) for this problem is fairly simple because the Hamiltonian does not depend on time and we can separate variables, writing

$$\frac{1}{2m} \left[\left(\frac{\partial S}{\partial q} \right)^2 + m^2 \omega^2 q^2 \right] = -\frac{\partial S}{\partial t}.$$

Integrating both sides gives us a function S having the form

$$S(\alpha, q, t) = W(\alpha, q) + V(\alpha, t),$$

where α is a constant. If such a separation is possible – and it often is possible – the partial differential equation can be solved reasonably easily. If such a separation cannot be carried out, the Hamilton–Jacobi technique is not a useful computational tool. As is often the case with partial differential equations, one *assumes* the separation is possible, then checks to see if the solution obtained obeys the differential equation and the boundary conditions.

Plugging the separated expression for S into the Hamilton–Jacobi equation we obtain

$$\frac{1}{2m}\left[\left(\frac{\partial W}{\partial q}\right)^2 + m^2\omega^2q^2\right] + \frac{\partial V}{\partial t} = 0,$$

or

$$\frac{1}{2m}\left[\left(\frac{\partial W}{\partial q}\right)^2 + m^2\omega^2q^2\right] = -\frac{\partial V}{\partial t}.$$

The left-hand side depends only on q and the right-hand side depends only on t. Since q and t are independent, both sides must be equal to the same constant. Let us denote this constant by α. Then,

$$V = -\alpha t,$$

and

$$\frac{1}{2m}\left[\left(\frac{dW}{dq_i}\right)^2 + m^2\omega^2q^2\right] = \alpha. \tag{6.9}$$

Note that the left-hand side of this equation is H so $\alpha = E$.

We have now expressed the Hamilton–Jacobi equation in time-independent form. The solution, W, is called "Hamilton's characteristic function."

For the problem at hand, we immediately obtain

$$W = \int dq\sqrt{2m\alpha - m^2\omega^2q^2},$$

and

$$S = -\alpha t + \int dq\sqrt{2m\alpha - m^2\omega^2q^2}.$$

By Equation (6.6)

$$\beta = \frac{\partial S}{\partial \alpha} = -t + \int dq\frac{2m}{\sqrt{2m\alpha - m^2\omega^2q^2}} = -t + \frac{1}{\omega}\sin^{-1}\left(q\sqrt{\frac{m\omega^2}{2\alpha}}\right),$$

so

$$q = \sqrt{\frac{2\alpha}{m\omega^2}}\sin\omega(t + \beta). \tag{6.10}$$

Similarly, by (6.8)

$$p = \frac{\partial S}{\partial q} = \sqrt{2m\alpha - m^2\omega^2q^2} = \sqrt{2m\alpha}\left(\sqrt{1 - \sin^2\omega(t + \beta)}\right),$$

or

$$p = \sqrt{2m\alpha} \cos \omega(t + \beta). \tag{6.11}$$

We have now obtained the transformation equations relating q and p at time t to the constant α and β. The last step is to determine α and β in terms of the initial values q_0 and p_0. This is easily done by setting $t = 0$ in Equations (6.10) and (6.11).

In this problem we had a single set of canonical variables so we did not need to subscript q. In general we would write $S(\alpha_i, q_i, t) = W(\alpha_i, q_i) + V(\alpha_i, t)$. The generalization of all the preceding equations to a system of many variables, $q_1, \ldots, q_n$ is straightforward.

Exercise 6.3 Determine p, q as functions of time for a free particle by solving the Hamilton–Jacobi equation.

6.3 Interpretation of Hamilton's principal function

We have treated Hamilton's principal function S simply as a generating function for a canonical transformation that takes us from q_i, p_i to a set of constants α_i, β_i which can be related to the initial conditions. However, it is instructive to note that since

$$S = S(q_1, \ldots, q_n; \alpha_1, \ldots, \alpha_n; t),$$

it follows that

$$\frac{dS}{dt} = \sum \frac{\partial S}{\partial q_i} \frac{dq_i}{dt} + \frac{\partial S}{\partial t},$$

(where we have used the fact that the α_i are constants). But $\frac{\partial S}{\partial q_i} = p_i$, so

$$\frac{dS}{dt} = \sum p_i \dot{q}_i + \frac{\partial S}{\partial t}.$$

Recall that the definition of the Hamiltonian is $H = \sum p_i \dot{q}_i - L$. Consequently,

$$\frac{dS}{dt} = H + L + \frac{\partial S}{\partial t}.$$

However, by Equation (6.2), $H + \frac{\partial S}{\partial t} = 0$, so

$$\frac{dS}{dt} = L,$$

or

$$S = \int L dt + \text{const.} \tag{6.12}$$

That is, Hamilton's principal function differs from the action integral by at most an additive constant.

6.4 Relationship to Schrödinger's equation

Louis de Broglie suggested that electrons behave like waves and hypothesized that the wavelength of these "matter waves" would be related to the momentum by $\lambda = h/p$, where h is Planck's constant. Schrödinger derived his well-known equation[1] from a consideration that if electrons behave as waves then they must obey some sort of wave equation, just as light waves obey the wave equation

$$\nabla^2 \phi - \frac{n^2}{c^2} \frac{\partial^2 \phi}{\partial t^2} = 0.$$

Now a possible solution to this equation is the plane wave solution in which the wave consists of plane wave fronts moving in (say) the z direction. This solution has the form

$$\phi = \phi_0 e^{ik(nz - ct)},$$

where k is the wave number. The quantity nz is called the optical path length or the "eikonal" and an expression of the form

$$\phi = \phi_0 e^{iZ} \tag{6.13}$$

is said to be in eikonal form. The quantity Z is the *phase* of the light wave. Note that an optical wavefront is a surface of constant Z.

If particles are waves, they can be represented by some sort of "wave function," but what would be the phase of such a wave function? It can be shown[2] that a surface of constant S propagates in phase space in precisely the same way as an ordinary wavefront propagates in configuration space. Thus it was reasonable for Schrödinger to select S, the action, to be the phase of the wave function of material particles. That is, he selected as the wavefunction for the electron the quantity

$$\Psi = \Psi_0 e^{iS/\hbar}, \tag{6.14}$$

[1] Much of the material in this section is based on the paper: Feynman's derivation of the Schrödinger equation by David Derkes, *Am. J. Phys*, **64**, pp. 881–884, July, 1996.

[2] H. Goldstein, *Classical Mechanics*, 2nd Edn, Section 10.8.

where $\hbar = h/2\pi$ makes the exponent dimensionless.

Now S obeys the Hamilton–Jacobi equation (6.4) which we can write as

$$\frac{\partial S}{\partial t} + H = \frac{\partial S}{\partial t} + \frac{p^2}{2m} + V = \frac{\partial S}{\partial t} + \frac{1}{2m}\left(\frac{\partial S}{\partial q}\right)^2 + V = 0. \qquad (6.15)$$

But note that if $\Psi = \Psi_0 e^{iS/\hbar}$, then

$$\frac{\partial \Psi}{\partial q} = \Psi_0 e^{iS/\hbar}\left(\frac{i}{\hbar}\frac{\partial S}{\partial q}\right) = \Psi\frac{i}{\hbar}\frac{\partial S}{\partial q},$$

so

$$\frac{\partial S}{\partial q} = -\frac{i\hbar}{\Psi}\frac{\partial \Psi}{\partial q}. \qquad (6.16)$$

Furthermore, by (6.2)

$$\frac{\partial S}{\partial t} = -H = -E,$$

so (6.15) becomes

$$-E + \frac{1}{2m}\left(-\frac{i\hbar}{\Psi}\frac{\partial \Psi}{\partial q}\right)\left(-\frac{i\hbar}{\Psi}\frac{\partial \Psi}{\partial q}\right)^* + V = 0, \qquad (6.17)$$

where we used the complex conjugate in evaluating the square because Ψ is a complex quantity.

Rearranging Equation (6.17) we obtain

$$(V - E)\Psi^*\Psi + \frac{\hbar^2}{2m}\left(\frac{\partial \Psi^*}{\partial q}\right)\left(\frac{\partial \Psi}{\partial q}\right) = 0. \qquad (6.18)$$

For convenience, let us define

$$M = (V - E)\Psi^*\Psi + \frac{\hbar^2}{2m}\left(\frac{\partial \Psi^*}{\partial q}\right)\left(\frac{\partial \Psi}{\partial q}\right).$$

Note that M is a function of Ψ, $\frac{\partial \Psi}{\partial q}$ and their complex conjugates. Schrödinger treated M as a Lagrangian with generalized coordinates Ψ, Ψ^*, $\frac{\partial \Psi}{\partial q}$, $\frac{\partial \Psi^*}{\partial q}$. This means that the integral $\int M dq$ is to be minimized, and this procedure leads to the following Euler–Lagrange equation for Ψ^*:

$$\frac{\partial M}{\partial \Psi^*} - \frac{\partial}{\partial q}\left(\frac{\partial M}{\partial\left(\frac{\partial \Psi^*}{\partial q}\right)}\right) = 0,$$

or

$$(V - E)\Psi - \frac{\hbar^2}{2m}\left(\frac{\partial^2 \Psi}{\partial q^2}\right) = 0,$$

or

$$-\frac{\hbar^2}{2m}\left(\frac{\partial^2 \Psi}{\partial q^2}\right) + V\Psi = E\Psi.$$

You will recognize this as the time-independent Schrödinger equation.

6.5 Problems

6.1 In polar coordinates the relativistic Kepler problem may be written as

$$H = c\left[p_r^2 + \frac{p_\phi^2}{r^2} + m^2 c^2\right]^{1/2} - \frac{k}{r}.$$

(a) Solve the Hamilton–Jacobi equation to find Hamilton's principal function (the action) $S(r, \phi, t)$. Leave your expressions in terms of integrals.
(b) From the action obtain a relation between ϕ and r of the form $\phi = \int f(r)dr + \phi_0$, where ϕ_0 is a constant.

6.2 In spherical coordinates the Hamiltonian for a particular system is given by

$$H = \frac{1}{2m}\left(p_r^2 + \frac{p_\theta^2}{r^2} + \frac{p_\phi^2}{r^2 \sin^2 \theta}\right) + a(r) + \frac{b(\theta)}{r^2}.$$

(a) Write the Hamilton–Jacobi equation. (b) Separate the variables. (c) Integrate to obtain S.

6.3 A charge q moves in the electric field generated by two fixed charges q_1 and q_2, separated by a distance $2d$. Let r_1 and r_2 be the distances from q to q_1 and q_2. Transform from cylindrical (ρ, ϕ, z) to elliptic coordinates (ξ, η, ϕ) defined by

$$\rho = d[(\xi^2 - 1)(1 - \eta^2)]^{1/2}$$
$$\phi = \phi$$
$$z = d\xi\eta.$$

(a) Show that $r_1 = d(\xi - \eta)$ and $r_2 = d(\xi + \eta)$. (b) Write the Lagrangian in elliptic coordinates. (c) Write the Hamiltonian in elliptic coordinates. (d) Obtain an expression for S.

6.4 A hoop of radius a lies in the xy plane and rotates with constant angular velocity ω about a vertical axis through a point on its rim as shown in Figure 6.1. A bead of mass m is free to slide on the hoop. Write the Hamiltonian in terms of p and ϕ, where p is the generalized momentum and ϕ is the angular position of the bead relative to the hoop diameter, as shown. (a) Obtain Hamilton's principal function. (b) Obtain an integral expression for $\phi(t)$.

6.5 We have expressed Hamilton's principal function as $S(q_i, \alpha_i, t)$. Show that p_i, q_i, are canonical variables, recalling that

$$p_i = \frac{\partial S}{\partial q_i},$$

$$q_i = q_i(\alpha, \beta, t),$$

where

$$\beta_i = Q_i = \frac{\partial S}{\partial \alpha_i}, \quad \text{and} \quad \alpha_i = P_i.$$

6.6 A particle of mass m is thrown vertically upward in a constant gravitational field. Solve the problem of the motion of the particle using the Hamilton–Jacobi method.

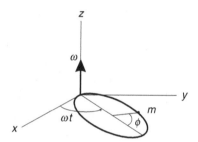

Figure 6.1   A bead on a hoop rotating at constant speed in the horizontal plane.

7

Continuous systems

So far, we have been studying discrete systems, that is, the mechanics of systems that are composed of a collection of *particles* (or rigid bodies that can be treated as particles). It is true that, from an atomistic point of view, all mechanical systems can be considered collections of particles. Nevertheless, from a computational point of view, it is much more convenient to consider some systems as continuous distributions of matter described by continuous properties, such as the mass density.

When applying variational methods to continuous systems it is convenient to introduce the Langrangian density $\mathcal{L}$ and the Hamiltonian density $\mathcal{H}$. These can be thought of as the Lagrangian or Hamiltonian per unit volume. The techniques we shall develop can be applied to a number of different continuous physical systems, and lead nicely into considerations of field theory. Although we mentioned it in Chapter 1, we have not emphasized the fact that the Lagrangian is defined to be a function that generates the equations of motion using the prescription

$$\frac{d}{dt}\frac{\partial L}{\partial \dot{q}} - \frac{\partial L}{\partial q} = 0$$

or an equivalent relation as determined by the Euler–Lagrange analysis. (That is, by applying Hamilton's principle.) The fact that $L = T - V$ for particle mechanics is secondary, although up to this point we have assumed that form for the Lagrangian.

7.1 A string

We begin our consideration with a very simple example of a continuous system, namely a string that is stretched between two end points (denoted

144

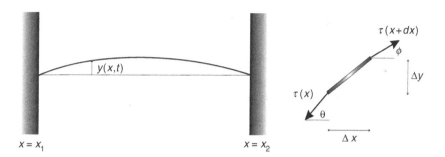

Figure 7.1   Left: an oscillating string. Right: an element of string at x subjected to tensions $\tau(x+dx)$ and $\tau(x)$.

x_1 and x_2). The tension in the string is τ. The mass per unit length of string is λ. You can imagine that the string has been plucked and a wave has been set up in it. See the left side of Figure 7.1.

At instant t the displacement of the string at location x is $y = y(x, t)$. Before treating this problem from a variational point of view, let us briefly go through the elementary derivation of the wave equation using Newton's second law. Consider an element of the string of length dx. Its mass is λdx. The forces acting on it are the tension on either end, that is, $\tau(x)$ and $\tau(x+dx)$, as indicated on the right side of Figure 7.1.

The figures are highly exaggerated, the string lies very close to the horizontal x axis. Therefore,

$$\sin\theta \simeq \tan\theta = \lim_{\Delta x \to 0} \frac{\Delta y}{\Delta x} = \frac{\partial y(x, t)}{\partial x}.$$

From $ma = F$ we have

$$\lambda(x)dx \frac{\partial^2 y}{\partial t^2} = \tau(x+dx)\sin\phi - \tau(x)\sin\theta$$

$$= \tau(x+dx)\frac{\partial y(x+dx, t)}{\partial x} - \tau(x)\frac{\partial y(x, t)}{\partial x}$$

$$= \tau(x)\frac{\partial y(x, t)}{\partial x} + dx\frac{\partial}{\partial x}\left(\tau(x)\frac{\partial y(x, t)}{\partial x}\right) + \cdots - \tau(x)\frac{\partial y(x, t)}{\partial x}$$

$$\lambda(x)dx \frac{\partial^2 y}{\partial t^2} = \frac{\partial}{\partial x}\left(\tau(x)\frac{\partial y}{\partial x}\right), \tag{7.1}$$

where we expanded the first term in a Taylor's series.

For simplicity, let us assume that λ and T are not functions of x. Then

$$\frac{\partial^2 y}{\partial t^2} = \frac{\tau}{\lambda}\frac{\partial^2 y}{\partial x^2},$$

which is the wave equation. The speed of the wave is $\sqrt{\lambda/\tau}$.

Thus we see that $F = ma$ leads us to the wave equation. But we are interested in obtaining this result using Lagrangian methods. I will write the appropriate Lagrangian for the continuous string by introducing a Lagrangian per unit length, or a Lagrangian density, which will be denoted $\mathcal{L}$. The Lagrangian for the system is

$$L = \int_{x_1}^{x_2} \mathcal{L} dx.$$

It is easy to appreciate that the kinetic energy density of an element of the string can be expressed as $\frac{1}{2}\lambda dx \dot{y}^2$, where λdx is the mass of the element of length dx, and $\dot{y}$ is its vertical velocity. It is not as obvious what to use for the potential energy. But note that the force $\frac{\partial}{\partial x}\left(\tau(x)\frac{\partial y}{\partial x}\right)$ is related to the potential energy through $F = -\frac{\partial V}{\partial x}$, so the potential energy can be written as $\frac{1}{2}\tau y'^2 dx$, where $y' = \partial y/\partial x$. Then the Lagrangian for the system can be written as

$$L = \int_{x_1}^{x_2} \mathcal{L} dx = \int_{x_1}^{x_2} \left[\frac{1}{2}\lambda(x)\dot{y}^2 - \frac{1}{2}\tau(x)y'^2 \right] dx.$$

If we make the reasonable assumption that the mass density λ and the tension τ are constants, we obtain

$$L = \frac{\lambda}{2}\int_{x_1}^{x_2} \dot{y}^2 dx - \frac{\tau}{2}\int_{x_1}^{x_2} y'^2 dx$$

and so, the Lagrangian density for this system is

$$\mathcal{L} = \frac{\lambda}{2}\dot{y}^2 - \frac{\tau}{2}y'^2.$$

This is often written in a more explicit form as

$$\mathcal{L} = \frac{\lambda}{2}\left(\frac{\partial y}{\partial t}\right)^2 - \frac{\tau}{2}\left(\frac{\partial y}{\partial x}\right)^2. \tag{7.2}$$

Exercise 7.1 Show that if the potential energy is given by $\frac{1}{2}\tau y'^2$, the force is given by the expression on the right-hand side of Equation (7.1).

Exercise 7.2 Consider a traveling wave in a string described in the usual way by $y(x, t) = A\cos(kx - \omega t)$. How are k and ω related to λ and τ?

Next let us determine the appropriate form for Lagrange's equations for this system. Going back to fundamentals, we recall that Lagrange's equation is a consequence of Hamilton's principle which we have stated as

$$\delta \int_{t_1}^{t_2} L\,dt = 0.$$

That is, the variation of the action taken between end points t_1 and t_2 is zero. But for our distributed one-dimensional system the Lagrangian may vary with position along the string. (It is obvious that the velocity of a mass point in the string depends on both t and x.) Assuming that the Lagrangian is distributed, the total Lagrangian of the system is

$$L = \int_{x_1}^{x_2} \mathcal{L}\,dx,$$

and Hamilton's principle then becomes

$$\delta \int_{t_1}^{t_2} \left[\int_{x_1}^{x_2} \mathcal{L}\,dx \right] dt = 0.$$

Let us consider the functional dependence of the Lagrangian density. It will be depend explicitly on position and time (x and t), as these are the independent variables. But we expect that it will depend implicitly on y, the displacement, $\dot{y} = \frac{\partial y}{\partial t}$, the velocity in the vertical direction, and $y' = \frac{\partial y}{\partial x}$, the "stretch." That is,

$$\mathcal{L} = \mathcal{L}(y, \dot{y}, y'; x, t).$$

Note that the independent variables x and t are the variables of integration in Hamilton's principle. The Lagrangian density at a particular place and time will, in general, depend on y, $\dot{y}$, and y'. We can consider y to be a generalized coordinate that specifies the configuration of the system. It will be varied in Hamilton's principle. Its variation at the end points must be zero. But now we have two sets of end points, both position and time. Note that we are not stating that y itself is zero at the boundaries, only that its *variation* at the boundaries is zero. That is,

$$\delta(y(t_1)) = \delta(y(t_2)) = 0,$$

and

$$\delta(y(x_1)) = \delta(y(x_2)) = 0.$$

This simply means that we require that all "paths" $y(x, t)$ begin at (x_1, t_1) and end at (x_2, t_2). For example, at the initial time, all locations on the varied string configuration correspond to the actual initial configuration, and at all times the varied string configuration has end points corresponding to the actual configuration (which happen to be $y(x_1, t) = 0$ and $y(x_2, t) = 0$).

Hamilton's principle is

$$0 = \delta \int_{t_1}^{t_2} L \, dt = \delta \int_{t_1}^{t_2} dt \int_{x_1}^{x_2} dx \mathcal{L}(y, y', \dot{y}; x, t)$$

$$0 = \int_{t_1}^{t_2} dt \int_{x_1}^{x_2} dx \left[\frac{\partial \mathcal{L}}{\partial y} \delta y + \frac{\partial \mathcal{L}}{\partial y'} \delta y' + \frac{\partial \mathcal{L}}{\partial \dot{y}} \delta \dot{y} \right]. \tag{7.3}$$

Note that

$$\delta y' = \delta \frac{\partial y}{\partial x} = \frac{\partial}{\partial x} \delta y, \qquad \text{where } t \text{ is held fixed,}$$

and

$$\delta \dot{y} = \delta \frac{\partial y}{\partial t} = \frac{\partial}{\partial t} \delta y, \qquad \text{where } x \text{ is held fixed.}$$

Consider the third term in Equation (7.3), keeping in mind that y (and $\dot{y}$) are values along the varied path (expressed, for example, as Y in Equation (2.5)):

$$\int_{t_1}^{t_2} dt \int_{x_1}^{x_2} dx \left[\frac{\partial \mathcal{L}}{\partial \dot{y}} \delta \dot{y} \right] = \int_{t_1}^{t_2} dt \int_{x_1}^{x_2} dx \left[\frac{\partial \mathcal{L}}{\partial \dot{y}} \frac{\partial \dot{y}}{\partial \epsilon} \delta \epsilon \right]$$

$$= \int_{x_1}^{x_2} dx \int_{t_1}^{t_2} dt \left[\frac{\partial \mathcal{L}}{\partial \dot{y}} \frac{\partial \dot{y}}{\partial \epsilon} \delta \epsilon \right]$$

$$= \int_{x_1}^{x_2} dx \int_{t_1}^{t_2} dt \frac{\partial \mathcal{L}}{\partial \dot{y}} \left(\frac{\partial}{\partial \epsilon} \frac{\partial y}{\partial t} \right) \delta \epsilon$$

$$= \int_{x_1}^{x_2} dx \int_{t_1}^{t_2} \frac{\partial \mathcal{L}}{\partial \dot{y}} \frac{\partial}{\partial t} \frac{\partial y}{\partial \epsilon} dt \delta \epsilon.$$

Integrating by parts

$$\int_{t_1}^{t_2} dt \int_{x_1}^{x_2} dx \left[\frac{\partial \mathcal{L}}{\partial \dot{y}} \delta \dot{y} \right] = \int_{x_1}^{x_2} dx \delta \epsilon \left\{ \left[\frac{\partial \mathcal{L}}{\partial \dot{y}} \frac{\partial y}{\partial \epsilon} \right]_{t_1}^{t_2} - \int_{t_1}^{t_2} \frac{\partial}{\partial t} \frac{\partial \mathcal{L}}{\partial \dot{y}} \frac{\partial y}{\partial \epsilon} dt \right\}$$

$$= \int_{x_1}^{x_2} dx \left\{ 0 - \int_{t_1}^{t_2} \frac{\partial}{\partial t} \frac{\partial \mathcal{L}}{\partial \dot{y}} \frac{\partial y}{\partial \epsilon} \delta \epsilon dt \right\}$$

$$= \int_{x_1}^{x_2} dx \left\{ - \int_{t_1}^{t_2} \frac{\partial}{\partial t} \frac{\partial \mathcal{L}}{\partial \dot{y}} \delta y dt \right\}$$

$$= - \int_{t_1}^{t_2} dt \int_{x_1}^{x_2} dx \left(\frac{\partial}{\partial t} \frac{\partial \mathcal{L}}{\partial \dot{y}} \right) \delta y.$$

Similarly, we can integrate the second term in Equation (7.3) by parts to obtain

$$\int_{t_1}^{t_2} dt \int_{x_1}^{x_2} dx \left[\frac{\partial \mathcal{L}}{\partial y'} \delta y' \right] = \int_{t_1}^{t_2} \int_{x_1}^{x_2} dt dx (-1) \frac{\partial}{\partial x} \frac{\partial \mathcal{L}}{\partial y'},$$

where we used the fact that the variation vanishes at the end points.

In conclusion, we appreciate that Hamilton's principle can be expressed as

$$0 = \int_{t_1}^{t_2} dt \int_{x_1}^{x_2} dx \left[\frac{\partial \mathcal{L}}{\partial y} - \frac{\partial}{\partial x} \frac{\partial \mathcal{L}}{\partial y'} - \frac{\partial}{\partial t} \frac{\partial \mathcal{L}}{\partial \dot{y}} \right] \delta y.$$

As usual, Hamilton's principle leads to the Lagrange equations. In the preceding equation we note that the variation δy is arbitrary so the term in brackets must be zero. Thus we obtain the general form for Lagrange's equations for a one-dimensional continuous system:

$$\frac{\partial}{\partial x} \frac{\partial \mathcal{L}}{\partial y'} + \frac{\partial}{\partial t} \frac{\partial \mathcal{L}}{\partial \dot{y}} - \frac{\partial \mathcal{L}}{\partial y} = 0. \tag{7.4}$$

We can write this in a more explicit but more complicated form as

$$\frac{\partial}{\partial x} \frac{\partial \mathcal{L}}{\partial \left(\frac{\partial y}{\partial x} \right)} + \frac{\partial}{\partial t} \frac{\partial \mathcal{L}}{\partial \left(\frac{\partial y}{\partial t} \right)} - \frac{\partial \mathcal{L}}{\partial y} = 0.$$

As a simple application, let us apply this form of the Lagrange equation to the Lagrangian density for a string which is

$$\mathcal{L} = \frac{\lambda}{2} \left(\frac{\partial y}{\partial t} \right)^2 - \frac{\tau}{2} \left(\frac{\partial y}{\partial x} \right)^2.$$

We appreciate that

$$\frac{\partial \mathcal{L}}{\partial \dot{y}} = \lambda \dot{y}, \qquad \frac{\partial \mathcal{L}}{\partial y'} = -\tau y', \qquad \text{and} \qquad \frac{\partial \mathcal{L}}{\partial y} = 0,$$

so that Equation (7.4) yields

$$\frac{\partial}{\partial t} (\lambda \dot{y}) + \frac{\partial}{\partial x} (-\tau y') = 0,$$

$$\lambda \ddot{y} - \tau y'' = 0.$$

That is,

$$\frac{\partial^2 y}{\partial t^2} = \frac{\tau}{\lambda} \frac{\partial^2 y}{\partial x^2},$$

the wave equation.

7.2 Generalization to three dimensions

We expressed the Lagrangian density for a one-dimensional system as

$$\mathcal{L} = \mathcal{L}(y, \dot{y}, y'; x, t),$$

or

$$\mathcal{L} = \mathcal{L}\left(y, \frac{\partial y}{\partial t}, \frac{\partial y}{\partial x}; x, t\right).$$

Generalizing to three dimensions, let the configuration of the system be specified by a scalar coordinate $w = w(x, y, z, t)$. We shall replace y with w and $\dot{y}$ with $\dot{w} = \frac{\partial w}{\partial t}$. This is straightforward. But now we need to replace y' with the derivatives of w with respect to x, y, z, specifically with $\frac{\partial w}{\partial x}, \frac{\partial w}{\partial y}, \frac{\partial w}{\partial z}$. Recalling the vector identity, $\nabla w = \hat{\mathbf{x}}\frac{\partial w}{\partial x} + \hat{\mathbf{y}}\frac{\partial w}{\partial y} + \hat{\mathbf{z}}\frac{\partial w}{\partial z}$ we see that a reasonably compact notation could be to let $\frac{\partial w}{\partial x}$ be represented as the x component of ∇w, thus,

$$\frac{\partial w}{\partial x} = \nabla_x w,$$

and similarly for the other two components. Then we write

$$\mathcal{L} = \mathcal{L}\left(w, \frac{\partial w}{\partial t}, \nabla w; x, y, z, t\right),$$

or

$$\mathcal{L} = \mathcal{L}\left(w, \frac{\partial w}{\partial t}, \nabla_x w, \nabla_y w, \nabla_z w; \mathbf{x}, t\right),$$

where $\mathbf{x}$ is the vector with components x, y, $z = x_1, x_2, x_3$.

The condition that the end points are fixed in time is

$$\delta w(t_1) = \delta w(t_2) = 0.$$

The Lagrange equation is, then,

$$\frac{\partial}{\partial t}\left(\frac{\partial \mathcal{L}}{\partial \left(\frac{\partial w}{\partial t}\right)}\right) + \sum_{i=1}^{3}\frac{\partial}{\partial x_i}\left(\frac{\partial \mathcal{L}}{\partial \left(\frac{\partial w}{\partial x_i}\right)}\right) - \frac{\partial \mathcal{L}}{\partial w} = 0, \qquad (7.5)$$

or, more compactly

$$\frac{\partial}{\partial t}\left(\frac{\partial \mathcal{L}}{\partial \dot{w}}\right) + \frac{\partial}{\partial x_i}\left(\frac{\partial \mathcal{L}}{\partial (\nabla_i w)}\right) - \frac{\partial \mathcal{L}}{\partial w} = 0, \qquad (7.6)$$

where the summation in the middle term has been replaced by the summation convention.

7.3 The Hamiltonian density

Returning to a one-dimensional system, we consider the Hamiltonian density $\mathcal{H}$, which bears the same relationship to the Lagrangian density $\mathcal{L}$ that the Hamiltonian H bears to the Lagrangian L, namely $H = p\dot{q} - L$.

For a one-dimensional continuous system we can write $\dot{Q}$ instead of $\dot{q}$, recalling that $\dot{q} = \dot{q}(t)$ is the generalized velocity of a *particle*, whereas $\dot{Q} = \dot{Q}(x, t)$ is a *distributed* quantity that depends on the entire range of positions.

To carry out our analogy between Hamiltonian density and Hamiltonian we need to define a distributed or continous variable analogous to the momentum. Based on the usual definition of generalized momentum as $p = \partial L / \partial \dot{q}$ we define the canonical momentum *density* by

$$\mathcal{P} = \frac{\partial \mathcal{L}}{\partial \dot{Q}}.$$

Then the one-dimensional Hamiltonian density is

$$\mathcal{H} = \mathcal{P}\dot{Q} - \mathcal{L}.$$

Let us now generalize to a three-dimensional system described by the configuration parameter Q. For convenience, let the Lagrange density not depend explicitly on t, (but it depends implicitly on time through the time dependence of Q). So,

$$\mathcal{L} = \mathcal{L}(Q, \dot{Q}, Q'_x, Q'_y, Q'_z; \mathbf{x}).$$

We are using a time-independent Lagrangian density because we want to explore the conservation of energy which involves investigating the partial derivative of $\mathcal{H}$ with respect to time:

$$\frac{\partial \mathcal{H}}{\partial t} = \frac{\partial}{\partial t} \left[\mathcal{P}\dot{Q} - \mathcal{L} \right]$$

$$\frac{\partial \mathcal{H}}{\partial t} = \frac{\partial \mathcal{P}}{\partial t} \frac{\partial Q}{\partial t} + \mathcal{P}\frac{\partial^2 Q}{\partial t^2} - \left[\frac{\partial \mathcal{L}}{\partial Q} \frac{\partial Q}{\partial t} + \frac{\partial \mathcal{L}}{\partial \dot{Q}} \frac{\partial \dot{Q}}{\partial t} + \sum_{k=1}^{3} \frac{\partial \mathcal{L}}{\partial Q'_k} \frac{\partial Q'_k}{\partial t} \right]. \quad (7.7)$$

But

$$\frac{\partial \mathcal{P}}{\partial t} = \frac{\partial}{\partial t} \frac{\partial \mathcal{L}}{\partial \dot{Q}},$$

and Lagrange's equation tells us that

$$\frac{\partial}{\partial t} \frac{\partial \mathcal{L}}{\partial \dot{Q}} = \frac{\partial \mathcal{L}}{\partial Q} - \sum_{k=1}^{3} \frac{\partial}{\partial x_k} \frac{\partial \mathcal{L}}{\partial Q'_k},$$

so

$$\frac{\partial \mathcal{P}}{\partial t} = \frac{\partial \mathcal{L}}{\partial Q} - \sum_{k=1}^{3} \frac{\partial}{\partial x_k} \frac{\partial \mathcal{L}}{\partial Q'_k}.$$

Plugging into Equation (7.7) we get

$$\frac{\partial \mathcal{H}}{\partial t} = \left[\frac{\partial \mathcal{L}}{\partial Q} - \sum_{k=1}^{3} \frac{\partial}{\partial x_k} \frac{\partial \mathcal{L}}{\partial Q'_k} \right] \frac{\partial Q}{\partial t} + \mathcal{P} \frac{\partial^2 Q}{\partial t^2} - \left[\frac{\partial \mathcal{L}}{\partial Q} \frac{\partial Q}{\partial t} + \frac{\partial \mathcal{L}}{\partial \dot{Q}} \frac{\partial \dot{Q}}{\partial t} \right. $$
$$\left. + \sum_{k=1}^{3} \frac{\partial \mathcal{L}}{\partial Q'_k} \frac{\partial}{\partial t} \frac{\partial Q}{\partial x_k} \right].$$

The first and fourth terms on the left cancel. Furthermore, the second term on the left can be expressed as:

$$\left[\sum_{k=1}^{3} \frac{\partial}{\partial x_k} \frac{\partial \mathcal{L}}{\partial Q'_k} \right] \frac{\partial Q}{\partial t} = \sum_{k=1}^{3} \frac{\partial}{\partial x_k} \left(\frac{\partial \mathcal{L}}{\partial Q'_k} \frac{\partial Q}{\partial t} \right) - \sum_{k=1}^{3} \frac{\partial \mathcal{L}}{\partial Q'_k} \frac{\partial}{\partial x_k} \frac{\partial Q}{\partial t},$$

leaving us with

$$\frac{\partial \mathcal{H}}{\partial t} = \mathcal{P} \frac{\partial^2 Q}{\partial t^2} - \sum_{k=1}^{3} \frac{\partial}{\partial x_k} \left(\frac{\partial \mathcal{L}}{\partial Q'_k} \frac{\partial Q}{\partial t} \right) - \frac{\partial \mathcal{L}}{\partial \dot{Q}} \frac{\partial \dot{Q}}{\partial t},$$

$$\frac{\partial \mathcal{H}}{\partial t} = - \sum_{k=1}^{3} \frac{\partial}{\partial x_k} \left(\frac{\partial \mathcal{L}}{\partial Q'_k} \frac{\partial Q}{\partial t} \right).$$

It is interesting to write the right-hand side as $-\nabla \cdot \vec{\mathcal{S}}$, where $\mathcal{S}_k = \left(\frac{\partial \mathcal{L}}{\partial Q'_k} \frac{\partial Q}{\partial t} \right)$. Then

$$\frac{\partial \mathcal{H}}{\partial t} + \nabla \cdot \vec{\mathcal{S}} = 0,$$

which we recognize as an equation of continuity, expressing the conservation of energy. This can be appreciated better by integrating over the volume of our system. Since $H = \int_V \mathcal{H} d^3 x$ we can write

$$\frac{dH}{dt} = \int_V \frac{\partial \mathcal{H}}{\partial t} d^3 x = - \int_V \nabla \cdot \vec{\mathcal{S}} d^3 x$$

and applying the divergence theorem we obtain

$$\frac{dH}{dt} = - \oint_A \vec{\mathcal{S}} \cdot \hat{n} da, \tag{7.8}$$

where A is the area of the surface. This tells us that a change in the energy contained in a volume is accompanied by a flow of energy across the surface bounding the volume. Thus $\vec{S}$ is an energy flux density.[1] If the surface is at infinity, the right hand side of Equation (7.8) is zero and the Hamiltonian is constant. An example might be an infinite system which is perturbed at some location. The Hamiltonian is also constant if $\vec{S}$ is perpendicular to the bounding surface so that $\vec{S} \cdot \hat{n} = 0$. An example would be a finite string with fixed end points.

Exercise 7.3 For a one-dimensional string, $\mathcal{L}$ is given by Equation (7.2). Show that $\vec{S}$ is given by $\vec{S} = -\tau \frac{\partial y}{\partial x} \frac{\partial y}{\partial t} \hat{x}$. Explain the direction of $\vec{S}$. Show that the energy flux is zero at the end points (x_1 and x_2).

Revisiting the string

If we apply the concepts of discrete mechanics to the string, we would write the relation between the Lagrange density, the kinetic energy density and the potential energy density in the form:

$$\mathcal{L} = \mathcal{T} - \mathcal{V}.$$

As we have seen, the kinetic energy density is

$$\mathcal{T} = \frac{1}{2}\lambda \left(\frac{\partial y}{\partial t}\right)^2 = \frac{1}{2}\lambda \dot{y}^2.$$

Since the potential energy does not depend on the velocity we have

$$\frac{\partial \mathcal{V}}{\partial \dot{y}} = 0$$

and consequently the generalized momentum density is

$$\mathcal{P} = \frac{\partial \mathcal{L}}{\partial \dot{y}} = \lambda \frac{\partial y}{\partial t}.$$

Then

$$\mathcal{H} = \mathcal{P}\dot{y} - \mathcal{L} = 2\mathcal{T} - (\mathcal{T} - \mathcal{V}) = \mathcal{T} + \mathcal{V}$$

as expected.

The Lagrangian density of the oscillating string is

$$\mathcal{L} = \frac{\lambda(x)}{2}\left(\frac{\partial y}{\partial t}\right)^2 - \frac{\tau(x)}{2}\left(\frac{\partial y}{\partial x}\right)^2 = \frac{\lambda(x)}{2}\dot{y}^2 - \frac{\tau(x)}{2}y'^2.$$

---

[1] Note the relationship to electromagnetism and the Poynting vector.

Denoting y by Q we can write

$$\mathcal{L} = \frac{1}{2}\lambda\dot{Q}^2 - \frac{1}{2}\tau Q'^2$$

$$\mathcal{P} = \frac{\partial\mathcal{L}}{\partial\dot{Q}} = \lambda(x)\dot{Q}.$$

Consequently,

$$\mathcal{H} = \mathcal{P}\dot{Q} - \mathcal{L},$$

and

$$\vec{\mathcal{S}} = \frac{\partial\mathcal{L}}{\partial Q'}\dot{Q} = -\tau(x)Q'\dot{Q}\hat{x}.$$

This is the energy flux vector, that is, the rate at which energy flows along the string.

Exercise 7.4 If $y = A\cos(kx - \omega t)$ (which is a traveling wave moving towards the right), show that

$$\mathcal{S}_x = A^2(\tau kc)\sin^2(kx - \omega t)$$

and that

$$\langle\mathcal{S}_x\rangle = \frac{1}{2}A^2\lambda c\omega^2,$$

where $c^2 = \lambda/\tau$ and $k = \omega/c$.

7.4 Another one-dimensional system

We now consider a different one-dimensional system. Although there are great similarities beween the two one-dimensional systems, we shall use the second system to elucidate some concepts that have not been brought out yet.

Imagine taking a hammer and hitting a large solid block of iron. A three-dimensional pressure/density wave will propagate through the solid. The perturbed molecules will be acted upon by intermolecular forces tending to bring them back to their equilibrium positions. We could, in principle, follow the motions of the individual molecules, but clearly it is advantageous to treat the iron block as a continuous system described by continuous parameters such as the density as a function of position and time.

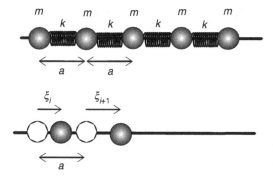

Figure 7.2 Beads of mass m on a thin, stiff wire, connected by springs of constant k. The equilibrium separation distance is a. The lower figure shows two beads displaced from equilibrium by distances ξ_i and ξ_{i+1}.

To better understand the physics of such a system it is convenient to begin with a one-dimensional system consisting of particles aligned in a straight line. Our "toy model" can be imagined as beads mounted on a thin rigid wire and connected to neighboring beads by springs, as illustrated in Figure 7.2.

Let each bead have mass m and each spring have spring constant k. Let the equilibrium spacing of the beads be a. If the equilibrium position of the ith bead is x_i, then its position when disturbed from equilibrium can be denoted $x_i + \xi_i$. In other words, ξ_i is the displacement from equilibrium of the ith particle. As indicated in the lower part of Figure 7.2 the distance between particle i and particle $i+1$ is $a + \xi_{i+1} - \xi_i$. The natural length of the springs is a, so the stretch of the spring is $(\xi_{i+1} - \xi_i)$ and the potential energy associated with the spring is $\frac{1}{2}k(\xi_{i+1} - \xi_i)^2$. The velocity of particle i relative to its equilibrium position (which is assumed to be at rest) is simply $\dot{\xi}_i$. Using these concepts we can write the Lagrangian in the usual form as

$$L = T - V = \sum_i \left(\frac{1}{2} m_i \dot{\xi}_i^2 - \frac{1}{2} k(\xi_{i+1} - \xi_i)^2 \right). \tag{7.9}$$

Exercise 7.5 Using the Lagrangian given in Equation (7.9), show that the equations of motion for this system of discrete particles are

$$m_i \ddot{\xi}_i - k(\xi_{i+1} - \xi_i) + k(\xi_i - \xi_{i-1}) = 0.$$

7.4.1 The limit of a continuous rod

We now represent the system of beads as a continuous system. We begin by
multiplying and dividing Equation (7.9) by a, thus:

$$L = a \sum_i \left[\frac{1}{2} \frac{m_i}{a} \dot{\xi}_i^2 - \frac{1}{2} ka \left(\frac{\xi_{i+1} - \xi_i}{a} \right)^2 \right].$$

As $a \to 0$, $m_i/a \to \lambda(x)$, the mass per unit length, i.e. the linear mass density
of the rod at location x. The term $\frac{\xi_{i+1} - \xi_i}{a}$ can be written as a continuous function also. Let $\xi_i = \xi(x)$ and $\xi_{i+1} = \xi(x + a)$. For convenience, let us write
Δx for a. Then

$$\frac{\xi_{i+1} - \xi_i}{a} \to \frac{\xi(x + \Delta x) - \xi(x)}{\Delta x}$$

and, in the limit $a \to 0$,

$$\frac{\xi_{i+1} - \xi_i}{a} \to \frac{d\xi}{dx}$$

and

$$ka \left(\frac{\xi_{i+1} - \xi_i}{a} \right)^2 \to k\Delta x \left(\frac{\partial \xi}{\partial x} \right)^2.$$

The summation becomes an integral over x, and

$$L = \frac{1}{2} \int \left[\lambda(x) \left(\frac{\partial \xi}{\partial t} \right)^2 - K \left(\frac{\partial \xi}{\partial x} \right)^2 \right] dx. \qquad (7.10)$$

Exercise 7.6 Show that K is Young's modulus.

Exercise 7.7 Show that the Lagrangian in Equation (7.10) yields the wave
equation.

Let us apply Hamilton's principle and derive Lagrange's equation. (This
derivation is similar to our previous considerations with the string.)

According to Hamilton's principle,

$$\delta \int_{t_1}^{t_2} \left[\int_{x_1}^{x_2} \mathcal{L} dx \right] dt = 0,$$

where $\mathcal{L} = \mathcal{L}(\xi, \dot{\xi}, \xi')$. For the problem at hand, the quantity being varied is ξ
and *not* x or t. Therefore, the variation of ξ at the end points is zero where the
end points are not only t_1 and t_2 but also x_1 and x_2.

When we express ξ as a continuous variable it is interpreted as representing the displacement from equilibrium of an infinitesimal segment of the rod. (This is difficult to visualize, so perhaps it is helpful to have a mental image of ξ as representing some other variable such as the density.)

To apply the variational method, we need to imagine a varied quantity $\xi(x, t, \epsilon)$ which differs infinitesimally from the "true" quantity $\xi(x, t, 0)$. That is, we define

$$\xi(x, t, \epsilon) = \xi(x, t, 0) + \epsilon \eta(x, t),$$

where $\xi(x, t, 0)$ is the "true" path, ϵ is a small parameter, and $\eta(x, t)$ is an arbitrary function that is zero at the end points. That is,

$$\eta(x_1, t_1) = 0, \qquad \text{and} \qquad \eta(x_2, t_2) = 0.$$

Following the usual prescription, we write

$$\frac{dI}{d\epsilon} = \int_{t_1}^{t_2} dt \int_{x_1}^{x_2} dx \left[\frac{\partial \mathcal{L}}{\partial \xi} \frac{\partial \xi}{\partial \epsilon} + \frac{\partial \mathcal{L}}{\partial \dot{\xi}} \frac{\partial \dot{\xi}}{\partial \epsilon} + \frac{\partial \mathcal{L}}{\partial \xi'} \frac{\partial \xi'}{\partial \epsilon} \right] = 0.$$

Integrating the second and third terms by parts we obtain

$$\int_{t_1}^{t_2} \frac{\partial \mathcal{L}}{\partial \dot{\xi}} \frac{\partial \dot{\xi}}{\partial \epsilon} dt = - \int_{t_1}^{t_2} \frac{\partial}{\partial t} \frac{\partial \mathcal{L}}{\partial \dot{\xi}} \frac{\partial \xi}{\partial \epsilon} dt,$$

and

$$\int_{x_1}^{x_2} \frac{\partial \mathcal{L}}{\partial \xi'} \frac{\partial \xi'}{\partial \epsilon} dx = - \int_{x_1}^{x_2} \frac{\partial}{\partial x} \frac{\partial \mathcal{L}}{\partial \xi'} \frac{\partial \xi}{\partial \epsilon} dx,$$

leading to

$$0 = \frac{dI}{d\epsilon} \bigg|_{\epsilon=0} = \int_{t_1}^{t_2} dt \int_{x_1}^{x_2} dx \left(\frac{\partial \mathcal{L}}{\partial \xi} - \frac{\partial}{\partial t} \frac{\partial \mathcal{L}}{\partial \dot{\xi}} - \frac{\partial}{\partial x} \frac{\partial \mathcal{L}}{\partial \xi'} \right) \left(\frac{\partial \xi}{\partial \epsilon} \bigg|_{\epsilon=0} \right).$$

Consequently, Lagrange's equations are

$$\frac{\partial}{\partial t} \left(\frac{\partial \mathcal{L}}{\partial \dot{\xi}} \right) + \frac{\partial}{\partial x} \left(\frac{\partial \mathcal{L}}{\partial \xi'} \right) - \frac{\partial \mathcal{L}}{\partial \xi} = 0, \tag{7.11}$$

as previously obtained (see Equation (7.4)).

It is interesting to note that for the discrete system we obtain one Lagrange equation for each bead, whereas for the continuous system we obtain a single Lagrange equation. On the other hand, we can consider that Equation (7.11) is an infinite number of equations, one for each of the infinite number of possible values of x.

Exercise 7.8 Show that Equation (7.11) applied to Equation (7.10) leads
to the wave equation.

Generalization to three dimensions

As previously noted, we can easily generate the relations developed above to
three dimensions. This would entail replacing the scalar displacement ξ by a
vector displacement, $\boldsymbol{\xi}$, with components ξ_i, where $i = (1, 2, 3)$ or (x, y, z).
The quantity $\boldsymbol{\xi}$ represents a physical quantity defined at every point in some
region of space, so it satisfies the definition of a *field*. In fact, we can let $\boldsymbol{\xi}$
represent some other physical quantity such as the pressure or velocity or electromagnetic field.

The field $\boldsymbol{\xi}$ is, consequently, defined at every point and at every instant of
time so $\boldsymbol{\xi} = \boldsymbol{\xi}(x, y, z, t)$ and the Lagrange equation is

$$\frac{\partial}{\partial t}\frac{\partial \mathcal{L}}{\partial \dot{\boldsymbol{\xi}}} + \frac{\partial}{\partial x}\left(\frac{\partial \mathcal{L}}{\partial (\partial \boldsymbol{\xi}/\partial x)}\right) + \frac{\partial}{\partial y}\left(\frac{\partial \mathcal{L}}{\partial (\partial \boldsymbol{\xi}/\partial y)}\right)$$

$$+ \frac{\partial}{\partial z}\left(\frac{\partial \mathcal{L}}{\partial (\partial \boldsymbol{\xi}/\partial z)}\right) - \frac{\partial \mathcal{L}}{\partial \boldsymbol{\xi}} = 0.$$

The ith component of this equation is

$$\frac{\partial}{\partial t}\frac{\partial \mathcal{L}}{\partial (\partial \xi_i/\partial t)} + \sum_{k=1}^{3}\frac{\partial}{\partial x_k}\frac{\partial \mathcal{L}}{\partial (\partial \xi_i/\partial x_k)} - \frac{\partial \mathcal{L}}{\partial \xi_i} = 0, \qquad (7.12)$$

where $x_k = x, y, z$. It might be noted that $\frac{\partial}{\partial t}$ and $\frac{\partial}{\partial x_k}$ are sometimes written as
the total derivatives $\frac{d}{dt}$ and $\frac{d}{dx_k}$ to indicate that during the differentiation one
must also include the implicit dependence of ξ_i on x_k or t.

It may be helpful to use a simpler notation, such as $\dot{\xi}_i$ for $\partial \xi_i/\partial t$ and ξ'_{ik} for
$\partial \xi_i/\partial x_k$. Then Equation (7.12) is written as

$$\frac{\partial}{\partial t}\frac{\partial \mathcal{L}}{\partial \dot{\boldsymbol{\xi}}} + \sum_{k=1}^{3}\frac{\partial}{\partial x_k}\frac{\partial \mathcal{L}}{\partial (\xi'_{ik})} - \frac{\partial \mathcal{L}}{\partial \xi_i} = 0. \qquad (7.13)$$

Simplified notation using the functional derivative

We can simplify the notation somewhat by using the functional derivative that
was defined by Equation (2.10) as

$$\frac{\delta \Phi}{\delta y} = \frac{\partial \Phi}{\partial y} - \frac{d}{dx}\frac{\partial \Phi}{\partial y'}.$$

For our purposes we write this as

$$\frac{\delta \mathcal{L}}{\delta \xi_i} = \frac{\partial \mathcal{L}}{\partial \xi_i} - \sum_{k=1}^{3} \frac{d}{dx_k} \frac{\partial \mathcal{L}}{\partial \nabla \xi_i},$$

or

$$\frac{\delta \mathcal{L}}{\delta \boldsymbol{\xi}} = \frac{\partial \mathcal{L}}{\partial \boldsymbol{\xi}} - \nabla \cdot \frac{\partial \mathcal{L}}{\partial \nabla \boldsymbol{\xi}}.$$

Consequently, in terms of the functional derivative, Hamilton's principle,

$$0 = \delta L = \int \sum_i \left(\frac{\partial \mathcal{L}}{\partial \xi_i} \delta \xi_i - \sum_{k=1}^{3} \frac{d}{dx_k} \frac{\partial \mathcal{L}}{\partial (\partial \xi_i / \partial x_k)} + \frac{\partial \mathcal{L}}{\partial \dot{\xi}_i} \delta \dot{\xi}_i \right) d^3 x,$$

can be expressed as

$$\delta L = \int \sum_i \left(\frac{\delta \mathcal{L}}{\delta \xi_i} \delta \xi_i + \frac{\partial \mathcal{L}}{\partial \dot{\xi}_i} \delta \dot{\xi}_i \right) d^3 x.$$

For the sake of simplicity, and to make the preceding more intelligible, let us go back to a one-dimensional system and a scalar variable, replacing ξ by Q. Furthermore, let us define the generalized momentum density in the usual way,

$$\mathcal{P} = \frac{\partial \mathcal{L}}{\partial \dot{Q}}.$$

Now

$$\mathcal{L} = \mathcal{L}(Q, \dot{Q}, Q'; x, t),$$

so

$$d\mathcal{L} = \frac{\partial \mathcal{L}}{\partial Q} dQ + \frac{\partial \mathcal{L}}{\partial \dot{Q}} d\dot{Q} + \frac{\partial \mathcal{L}}{\partial Q'} dQ' + \frac{\partial \mathcal{L}}{\partial t} dt.$$

Lagrange's equation (Equation (7.11)) can be written (replacing ξ with Q) as

$$\frac{\partial}{\partial t} \frac{\partial \mathcal{L}}{\partial \dot{Q}} + \frac{\partial}{\partial x} \frac{\partial \mathcal{L}}{\partial Q'} - \frac{\partial \mathcal{L}}{\partial Q} = 0.$$

The one-dimensional functional derivative can be written as

$$\frac{\delta \mathcal{L}}{\delta Q} = \frac{\partial \mathcal{L}}{\partial Q} - \frac{\partial}{\partial x} \frac{\partial \mathcal{L}}{\partial Q'}.$$

Similarly,

$$\frac{\delta \mathcal{L}}{\delta \dot{Q}} = \frac{\partial \mathcal{L}}{\partial \dot{Q}} - \frac{\partial}{\partial x} \frac{\partial \mathcal{L}}{\partial \dot{Q}} = \frac{\partial \mathcal{L}}{\partial \dot{Q}} - \frac{\partial}{\partial x} \frac{\partial \mathcal{L}}{\partial (\partial \dot{Q}/\partial x)} = \frac{\partial \mathcal{L}}{\partial \dot{Q}},$$

where we set $\frac{\partial \mathcal{L}}{\partial(\partial \dot{Q}/\partial x)}$ to zero because $\mathcal{L}$ does not depend on the derivative of the velocity with respect to position. Therefore, we obtain the following form for the Lagrange equation

$$\frac{\partial}{\partial t}\frac{\delta \mathcal{L}}{\delta \dot{Q}} - \frac{\delta \mathcal{L}}{\delta Q} = 0. \tag{7.14}$$

Thus we have obtained Lagrange's equation in the well-known form, except we are expressing it in terms of functional derivatives rather than partial derivatives.

We note also that the functional derivative of $\mathcal{L}$ with respect to $\dot{\mathcal{P}}$ is

$$\frac{\delta \mathcal{L}}{\delta \dot{\mathcal{P}}} = \frac{\partial \mathcal{L}}{\partial \dot{\mathcal{P}}} - \frac{\partial}{\partial x}\frac{\partial \mathcal{L}}{\partial(\partial \dot{\mathcal{P}}/\partial x)}.$$

Again the last term is zero so

$$\frac{\delta \mathcal{L}}{\delta \dot{\mathcal{P}}} = \frac{\partial \mathcal{L}}{\partial \dot{\mathcal{P}}},$$

and consequently

$$\frac{\delta \mathcal{L}}{\delta Q} = \frac{\partial}{\partial t}\frac{\delta \mathcal{L}}{\delta \dot{Q}} = \frac{\partial}{\partial t}\mathcal{P} = \dot{\mathcal{P}}. \tag{7.15}$$

Exercise 7.9 Show that the functional derivative of $\Theta(\rho) = c_F \int (\rho(\mathbf{r}))^{5/3}\, dr$ is $\frac{5}{3}c_F(\rho(\mathbf{r}))^{2/3}$. The quantity c_F is a constant.

Exercise 7.10 Show that

$$\frac{\delta \mathcal{L}}{\delta \dot{\xi}} = \frac{\partial \mathcal{L}}{\partial \dot{\xi}}.$$

7.4.2 The continuous Hamiltonian and the canonical field equations

For a one-dimensional system, recalling that we have defined the continuous generalized momentum by

$$\mathcal{P} = \frac{\partial \mathcal{L}}{\partial \dot{Q}},$$

we can write the Hamiltonian density as

$$\mathcal{H} = \mathcal{P}\dot{Q} - \mathcal{L}.$$

For a three-dimensional system we write

$$\mathcal{H} = \sum_{i=1}^{3} \mathcal{P}_i \dot{Q}_i - \mathcal{L}.$$

Consequently,

$$H = \int_V \mathcal{H} d^3 x = \int_V \sum_{i=1}^{3} \left(\mathcal{P}_i \dot{Q}_i - \mathcal{L} \right) d^3 x.$$

Note that

$$\mathcal{P}_i \equiv \frac{\partial \mathcal{L}}{\partial \dot{Q}_i} = \frac{\delta \mathcal{L}}{\delta \dot{Q}_i}$$

because $\mathcal{L}$ does not depend on $\partial \dot{Q}_i / \partial x_j$.

Now the coordinates Q_i and the momenta $\mathcal{P}_i$ are functions of the independent parameters (x, y, z, t). That is,

$$\mathcal{H} = \mathcal{H}(\mathcal{P}_i, Q_i, \partial Q_i / \partial x_j; \mathbf{x}, t).$$

Therefore,

$$dH = d \left[\int_V \mathcal{H} d^3 x \right]$$

$$= \int_V \left(\sum_i \frac{\partial \mathcal{H}}{\partial \mathcal{P}_i} d\mathcal{P}_i + \frac{\partial \mathcal{H}}{\partial Q_i} dQ_i + \sum_{k=1}^{3} \frac{\partial \mathcal{H}}{\partial (\partial Q_i / \partial x_j)} d \left(\frac{\partial Q_i}{\partial x_k} \right) + \frac{\partial \mathcal{H}}{\partial t} dt \right) d^3 x.$$

In terms of the functional derivative this can be expressed as

$$dH = \int_V \left[\sum_i \left(\frac{\delta \mathcal{H}}{\delta \mathcal{P}_i} d\mathcal{P}_i + \frac{\delta \mathcal{H}}{\delta Q_i} dQ_i \right) + \frac{\partial \mathcal{H}}{\partial t} dt \right] d^3 x. \qquad (7.16)$$

But recall that

$$H = \int_V \sum_i \left(\mathcal{P}_i \dot{Q}_i - \mathcal{L} \right) d^3 x,$$

so we can write dH as

$$dH = \int_V \left\{ \sum_i \left(\mathcal{P}_i d\dot{Q}_i + \dot{Q}_i d\mathcal{P}_i - \frac{\delta \mathcal{L}}{\delta Q_i} dQ_i - \frac{\delta \mathcal{L}}{\delta \dot{Q}_i} d\dot{Q}_i \right) - \frac{\partial \mathcal{L}}{\partial t} dt \right\} d^3 x.$$

By the definition of $\mathcal{P}_i$, the first and fourth terms on the right-hand side cancel, leaving us with

$$dH = \int_V \left\{ \sum_i \left(\dot{Q}_i d\mathcal{P}_i - \frac{\delta \mathcal{L}}{\delta Q_i} dQ_i \right) - \frac{\partial \mathcal{L}}{\partial t} dt \right\} d^3 x.$$

But by Equation (7.15), $\delta\mathcal{L}/\delta Q_i = \dot{\mathcal{P}}_i$, so

$$dH = \int_V \left\{ \sum_i \left(-\dot{\mathcal{P}}_i dQ_i + \dot{Q}_i d\mathcal{P}_i \right) - \frac{\partial \mathcal{L}}{\partial t} dt \right\} d^3x. \qquad (7.17)$$

Equating terms in Equations (7.16) and (7.17) we obtain

$$\frac{\delta\mathcal{H}}{\delta Q_i} = -\dot{\mathcal{P}}_i \quad \text{and} \quad \frac{\delta\mathcal{H}}{\delta\mathcal{P}_i} = \dot{Q}_i \quad \text{and} \quad \frac{\partial\mathcal{H}}{\partial t} = -\frac{\partial\mathcal{L}}{\partial t}.$$

These are the continuous versions of Hamilton's canonical equations.

Exercise 7.11 Show that Equation (7.16) is correctly expressed in terms of the functional derivative.

Exercise 7.12 Show that in terms of ordinary partial derivatives, Hamilton's canonical equations for continuous systems lose their symmetry, taking on the form

$$\frac{\partial\mathcal{H}}{\partial Q_i} - \sum_{j=1}^{3} \frac{d}{dx_j} \frac{\partial\mathcal{H}}{\partial(\partial Q_i/\partial x_j)} = -\dot{\mathcal{P}}_i$$

$$\frac{\partial\mathcal{H}}{\partial\mathcal{P}_i} = \dot{Q}_i.$$

7.5 The electromagnetic field

An interesting application of the formalism for continuous systems is the treatment of electromagnetic fields.

Recall that by definition a field is a physical quantity that is defined at every point in some region of space. The electromagnetic field is usually described by Maxwell's equations which are expressions for the divergence and curl of the electric and magnetic fields.[2] In vacuum these are given by

$$\nabla \cdot \mathbf{E} = \rho/\epsilon_0 \qquad\qquad \nabla \cdot \mathbf{B} = 0 \qquad\qquad (7.18)$$

$$\nabla \times \mathbf{E} = -\frac{\partial \mathbf{B}}{\partial t} \qquad \nabla \times \mathbf{B} = \mu_0 \mathbf{J} + \epsilon_0\mu_0 \frac{\partial \mathbf{E}}{\partial t}.$$

Here ρ is the charge density, and $\mathbf{J}$ is the current density. The constants ϵ_0 and μ_0 are the permittivity and permeability of free space and are included in

---

[2] The Helmholtz theorem tells us that if we know the divergence and curl of a vector field, then we can determine the field itself. For this reason the divergence and curl of a field are often called the "sources" of the field. Maxwell's equations are simply the source equations for the electric and magnetic fields. These equations tell us that the sources of the electromagnetic fields are charge densities and current densities.

the equations because we are using S.I. units (sometimes called "rationalized MKS units").

Now it turns out that **E** and **B** are not independent, as is clear from Maxwell's equations, and it further turns out that two of them can be derived from the other two, as we shall show in a moment.

The fields **E** and **B** can be obtained from the "potentials" ϕ and **A**, which are called the "scalar potential" and the "vector potential". The expressions for **E** and **B** in terms of the potentials are

$$\mathbf{E} = -\nabla\phi - \frac{\partial \mathbf{A}}{\partial t}, \tag{7.19}$$

$$\mathbf{B} = \nabla \times \mathbf{A}.$$

As mentioned previously, the Lagrangian is a function that can be used to generate the equations of motion. In this case, the "equations of motion" are the field equations, that is, Maxwell's equations. The independent generalized coordinates will be the scalar potential ϕ and the components of the vector potential **A**, that is, A_x, A_y, A_z.

The Lagrangian density that will generate the Maxwell equations is

$$\mathcal{L} = \frac{1}{2}\left(\epsilon_0 E^2 + \frac{1}{\mu_0}B^2\right) - \rho\phi + \mathbf{J} \cdot \mathbf{A}, \tag{7.20}$$

where E^2 and B^2 are to be expressed in terms of the potentials, using Equation (7.19). Note that this Lagrangian density is made up of (1) the energy density in the electromagnetic fields, (2) the energy due to the interaction of the charge density ρ with the scalar potential ϕ and (3) the interaction of the current density **J** with the vector potential **A**. To determine the field equations we use the Lagrange equation in the form

$$\frac{d}{dt}\frac{\partial \mathcal{L}}{\partial \dot{Q}_i} + \sum_{j=1}^{3}\frac{d}{dx_j}\left(\frac{\partial \mathcal{L}}{\partial(\partial Q_i/\partial x_j)}\right) - \frac{\partial \mathcal{L}}{\partial Q_i} = 0,$$

where Q_i is ϕ or one of the three components of **A**. We now show that evaluating this Lagrange equation leads to the Maxwell equations. Let us begin by setting Q_i equal to ϕ because that is the easiest case. The Lagrange equation becomes

$$\frac{d}{dt}\frac{\partial \mathcal{L}}{\partial \dot{\phi}} + \sum_{j=1}^{3}\frac{d}{dx_j}\left(\frac{\partial \mathcal{L}}{\partial(\partial\phi/\partial x_j)}\right) - \frac{\partial \mathcal{L}}{\partial \phi} = 0. \tag{7.21}$$

From Equations (7.20) and (7.19) we appreciate that the Lagrangian density does not depend on $\dot{\phi}$. It does, however, depend on $\dot{A}_i$ through the dependence of $\mathbf{E}$ on $\frac{\partial \mathbf{A}}{\partial t}$. It also depends on the spatial derivatives of ϕ and $\mathbf{A}$ through $\nabla \phi$ and $\nabla \times \mathbf{A}$.

Nevertheless, $\partial \mathcal{L} / \partial \phi = 0$ so we can discard the first term in Equation (7.21). The second term is

$$T_2 = \frac{d}{dx}\left(\frac{\partial \mathcal{L}}{\partial(\partial \phi / \partial x)}\right) + \frac{d}{dy}\left(\frac{\partial \mathcal{L}}{\partial(\partial \phi / \partial y)}\right) + \frac{d}{dz}\left(\frac{\partial \mathcal{L}}{\partial(\partial \phi / \partial z)}\right).$$

$\mathbf{B}$ does not depend on ϕ and $\mathbf{E}$ depends on the spatial derivatives of ϕ according to

$$\mathbf{E} = -\nabla \phi = -\left(\hat{\mathbf{x}}\frac{\partial \phi}{\partial x} + \hat{\mathbf{y}}\frac{\partial \phi}{\partial y} + \hat{\mathbf{z}}\frac{\partial \phi}{\partial z}\right),$$

so

$$-\frac{\partial \phi}{\partial x} = E_x, \qquad -\frac{\partial \phi}{\partial y} = E_y, \qquad -\frac{\partial \phi}{\partial z} = E_z.$$

Therefore

$$T_2 = -\frac{d}{dx}\frac{\partial \mathcal{L}}{\partial E_x} - \frac{d}{dy}\frac{\partial \mathcal{L}}{\partial E_y} - \frac{d}{dz}\frac{\partial \mathcal{L}}{\partial E_z}$$

$$= -\nabla \cdot \left(\frac{\partial \mathcal{L}}{\partial E_x}\hat{\mathbf{x}} + \frac{\partial \mathcal{L}}{\partial E_y}\hat{\mathbf{y}} + \frac{\partial \mathcal{L}}{\partial E_z}\hat{\mathbf{z}}\right)$$

$$= -\epsilon_0 \nabla \cdot \mathbf{E},$$

because

$$\frac{\partial \mathcal{L}}{\partial E_x} = \frac{\partial \left(\frac{1}{2}\epsilon_0 E^2\right)}{\partial E_x} = \frac{\partial \left(\frac{1}{2}\epsilon_0 \left(E_x^2 + E_y^2 + E_z^2\right)\right)}{\partial E_x} = \epsilon_0 E_x.$$

Finally, the third term is

$$T_3 = -\frac{\partial \mathcal{L}}{\partial \phi} = +\rho.$$

Consequently,

$$T_1 + T_2 + T_3 = 0 - \epsilon_0 \nabla \cdot \mathbf{E} + \rho = 0,$$

so

$$\nabla \cdot \mathbf{E} = \rho / \epsilon_0.$$

Thus we have derived the first of the Maxwell equations.

Another Maxwell equation is obtained as follows. Let the continuous generalized coordinate be A_x, the x component of the vector potential. Note that

$$\mathbf{E} = -\nabla \phi - \frac{\partial \mathbf{A}}{\partial t}.$$

Therefore

$$\mathbf{E} = E_x\hat{\mathbf{x}} + E_y\hat{\mathbf{y}} + E_z\hat{\mathbf{z}} = \left(-\frac{\partial\phi}{\partial x} - \dot{A}_x\right)\hat{\mathbf{x}} + \left(-\frac{\partial\phi}{\partial x} - \dot{A}_y\right)\hat{\mathbf{y}} + \left(-\frac{\partial\phi}{\partial z} - \dot{A}_z\right)\hat{\mathbf{z}}.$$

Similarly, $\mathbf{B} = \nabla \times \mathbf{A}$, therefore

$$\mathbf{B} = \left(A'_{zy} - A'_{yz}\right)\hat{\mathbf{x}} + \left(A'_{xz} - A'_{zx}\right)\hat{\mathbf{y}} + \left(A'_{yx} - A'_{xy}\right)\hat{\mathbf{z}}.$$

The Lagrange equation is

$$\frac{d}{dt}\frac{\partial\mathcal{L}}{\partial\dot{A}_x} + \sum_{j=1}^{3}\frac{d}{dx_j}\left(\frac{\partial\mathcal{L}}{\partial A'_{xj}}\right) - \frac{\partial\mathcal{L}}{\partial A_x} = 0.$$

Since the only term in $\mathcal{L}$ that depends on $\dot{A}_x$ is $\epsilon_0 E^2/2$ and the only term depending on A' is $-B^2/2\mu_0$, and the only term depending on A_x is $\mathbf{J}\cdot\mathbf{A}$, let us write

$$0 = \frac{d}{dt}\frac{\partial\epsilon_0 E^2/2}{\partial\dot{A}_x} + \sum_{j=1}^{3}\frac{d}{dx_j}\left(-\frac{\partial B^2/2\mu_0}{\partial A'_{xj}}\right) - \frac{\partial\left(\mathbf{J}\cdot\mathbf{A}\right)}{\partial A_x}$$

$$0 = \frac{d}{dt}\frac{\epsilon_0}{2}\frac{\partial E^2}{\partial E}\frac{\partial E}{\partial\dot{A}_x} - \sum_{j=1}^{3}\frac{d}{dx_j}\frac{1}{2\mu_0}\frac{\partial B^2}{\partial B}\frac{\partial B}{\partial A'_{xj}} - \frac{\partial}{\partial A_z}\left(J_x A_x + J_y A_y + J_z A_z\right)$$

$$0 = -\epsilon_0\frac{dE_x}{dt} + \frac{1}{\mu_0}\left(\frac{\partial B_z}{\partial y} - \frac{\partial B_y}{\partial z}\right) - J_x$$

$$0 = -\epsilon_0\frac{dE_x}{dt} + \frac{1}{\mu_0}(\nabla\times\mathbf{B})_x - J_x,$$

or

$$(\nabla\times\mathbf{B})_x = \epsilon_0\mu_0\frac{\partial E_x}{\partial t} + \mu_0 J_x.$$

Similarly, the other two components of $\mathbf{A}$ lead to the other two components of the last Maxwell equation in (7.18).

Thus we see that the continuous formulation of the Lagrangian leads to two of the Maxwell equations. But what about the other two Maxwell equations? These can be obtained from the two we have derived. Specifically, since $\mathbf{B} = \nabla\times\mathbf{A}$ then

$$\nabla\cdot\mathbf{B} = \text{div}(\mathbf{B}) = \text{div}(\nabla\times\mathbf{A}) = \text{div}\,\text{curl}(\mathbf{A}) = 0$$

because the divergence of the curl of any vector function is zero. Similarly, since $\mathbf{E} = -\nabla\phi - \partial\mathbf{A}/\partial t$,

$$\nabla\times\mathbf{E} = \text{curl}(\mathbf{E}) = -\text{curl}\,\text{grad}\phi - \frac{\partial}{\partial t}(\text{curl}\,\mathbf{A}) = -\frac{\partial\mathbf{B}}{\partial t}$$

because the curl of the gradient of any scalar function is zero.

A difficulty with our description of the electromagnetic field is the fact that there is no term in the electromagnetic field energy that is equivalent to a kinetic energy, so we cannot define a generalized momentum and define a Hamiltonian density.

7.6 Conclusion

In this short chapter we have illustrated the procedure for a Lagrangian approach to continuous systems. I hope that you have appreciated that the main difficulty in our development has been the cumbersome notation. If you can get past the complicated looking equations, you will appreciate that the essential concepts are not really different than the concepts we have been using for discrete systems.

We have not only obtained Lagrange's equation for continuous fields, we have also developed the Hamiltonian and Hamilton's canonical equations of motion. Finally, we have seen that the Lagrangian density $\mathcal{L}$ describes the electromagnetic field.

The theory outlined in this chapter is applicable in several other fields such as fluid dynamics and is particularly important in the development of the quantum theory of fields.

7.7 Problems

7.1 Show that adding $\partial \mathcal{L}/\partial t + \partial \mathcal{L}/\partial x$ to a one-dimensional Lagrangian density does not change the Lagrange equation.

7.2 Consider the Lagrangian density

$$\mathcal{L} = \frac{i\hbar}{2}(\Psi^* \dot{\Psi} - \dot{\Psi}^* \Psi) - \frac{\hbar^2}{2m}\nabla\Psi^* \cdot \nabla\Psi - V(r)\Psi^*\Psi.$$

Show that $-\delta\mathcal{L}/\delta\Psi^*$ generates that Schrödinger equation.

7.3 Consider the Lagrangian density

$$\mathcal{L} = \lambda \left[\frac{1}{2}\left(\frac{\partial \eta}{\partial t}\right)^2 - \frac{1}{2}v^2\left(\frac{\partial \eta}{\partial x}\right)^2 - \frac{1}{2}\Omega\eta^2 \right],$$

where $\lambda = $ linear mass density, $v = $ wave velocity and Ω^2 is a constant. Show that the Lagrange equation generates the one-dimensional Klein–Gordon equation:

$$\frac{\partial^2 \eta}{\partial t^2} - v^2\frac{\partial^2 \eta}{\partial x^2} + \Omega^2\eta.$$

7.4 The Lagrangian density for irrotational isentropic fluid flow is

$$\mathcal{L} = \rho \frac{\partial \Phi}{\partial t} - \frac{1}{2}\rho (\nabla \Phi)^2 - \rho U - \rho \varepsilon(\rho),$$

where Φ is the velocity potential ($\mathbf{v} = -\nabla \Phi$), ρ is the density, ε is the internal energy density and U is the potential energy per unit mass. The Lagrangian density is a function of Φ and ρ, which can be considered as the generalized fields. (a) Show that the generalized momentum density is ρ. (b) Show that the equation of motion for the generalized momentum density is the equation of continuity for the fluid.

7.5 The "sound field" is made up of longitudinal vibrations in a gas. Is the displacement of an element of the gas is described by $\boldsymbol{\eta}$ (with components $\eta_i = \eta_x, \eta_y, \eta_z$) the kinetic energy density is

$$\mathcal{T} = \frac{1}{2}\rho_0(\dot{\eta}_x^2 + \dot{\eta}_y^2 + \dot{\eta}_z^2),$$

where ρ_0 is the equilibrium value of the mass density. The potential energy density is

$$\mathcal{V} = -P_0 \nabla \cdot \boldsymbol{\eta} + \frac{1}{2}\gamma P_0 (\nabla \cdot \boldsymbol{\eta})^2 ,$$

where P_0 is the equilibrium value of the pressure and γ is the ratio of specific heat at constant pressure to specific heat at constant volume. (a) Write the Lagrangian density. (b) Obtain the equations of motion. (c) Explain why the term $P_0 \nabla \cdot \boldsymbol{\eta}$ does not contribute to the equation of motion. (d) Combine the equations of motion to obtain a three dimensional wave equation.

7.6 Consider a function $F = F(q_i, p_i)$ that can be expressed in terms of a density function $\mathcal{F}$, so that $F = \int_V \mathcal{F} d^3 x$. Assume that $\mathcal{F}$ does not depend explicitly on time. (a) Show that the time derivative of F can be expressed as

$$\frac{dF}{dt} = \int_V \sum_i \left(\frac{\delta \mathcal{F}}{\delta Q_i} \dot{Q}_i + \frac{\delta \mathcal{F}}{\partial \mathcal{P}} \dot{\mathcal{P}} \right) d^3 x.$$

(b) Apply the canonical equations and obtain an analog of the Poisson bracket.

Bibliography

Alexander L. Fetter and John Dirk Walecka, *Theoretical Mechanics of Particles and Continua,* McGraw-Hill, New York, 1980.

Herbert Goldstein, *Classical Mechanics,* Addison-Wesley Pub. Co., Reading MA, USA, 1950.

Herbert Goldstein, *Classical Mechanics, 2nd Edn,* Addison-Wesley Pub. Co., Reading MA, USA, 1980.

Louis N. Hand and Janet D. Finch, *Analytical Mechanics*, Cambridge University Press, 1998.

Jorge V. Jose and Eugene J. Saletan, *Classical Dynamics, A Contemporary Approach*, Cambridge University Press, 1998.

L. D. Landau and E. M. Lifshitz, *Mechanics, Vol 1 of A Course of Theoretical Physics*, Pergamon Press, Oxford, 1976.

Cornelius Lanczos, *The Variational Principles of Mechanics*, The University of Toronto Press, 1970. Reprinted by Dover Press, New York, 1986.

K. F. Riley, M. P. Hobson and S. J. Bence, *Mathematical Methods for Physics and Engineering, 2nd Edn*, Cambridge University Press, 2002.

Stephen T. Thornton and Jerry B. Marion, *Classical Dynamics of Particles and Systems*, Brooks/Cole, Belmont CA, 2004.

Answers to selected problems

1.1 $t_y < \tau + v_0/2a_c$

1.2 $t = (2l\frac{M+3m}{mg})^{1/2}$

1.5 $\ddot{s} = \frac{2}{3}g \sin a$

1.6 $L = \frac{1}{2}(M+m)\dot{x}^2 + Ml\dot{x}\dot{\theta}\cos\theta + \frac{1}{2}Ml^2\dot{\theta}^2 + mgl\cos\theta$

2.2 $\theta = \alpha$, $r\sin\alpha = \frac{c_2}{\cos(\phi\sin\alpha + c_1)}$

2.3 Intersection of $r = 1 + \cos\theta$ with $z = a + b\sin(\theta/2)$.

2.4 Right circular cylinder.

2.7 Answer: $\rho = k\cosh(z/k)$.

2.12 Arc of a circle

2.13 $R = \frac{1}{2}H$

3.4 Suggestion: Find two "end points" for the path, defined by two times, for an arbitrary initial velocity (v_{0x}, v_{0y}). Then define a family of parabolas (or some other curves) that go through those end points.

3.5 $y = a\cosh\frac{x-b}{a}$

3.6 $N = mg\cos\theta - m(R_1 + R_2)\dot{\theta}^2$; $\theta = 39.72°$

3.7 $F = (mg\cos\theta)\left(\frac{M}{M+m\sin^2\theta}\right)$

3.8 $L = \frac{1}{2}m(2\dot{l}_1^2 + l_1^2\dot{\theta}^2) - mgl_1 +$ const where l is distance from m_1 to hole.

4.1 $H = \left(\frac{p_{\theta_1}^2 m_2 l_2^2 + p_{\theta_2}^2(m_1+m_2)l_1^2 - 2p_{\theta_1}p_{\theta_1}m_2l_1l_2\cos(\theta_1-\theta_2)}{2l_1^2 l_2^2 m_2(m_1+m_2\sin^2(\theta_1-\theta_2))}\right) -$
$(m_1 + m_2)gl_1\cos\theta_1 - m_2gl_2\cos\theta_2$

4.3 $H = \frac{p_\theta^2}{2ml^2} - \frac{p_\theta a\omega}{l}\cos(\omega t + \theta) + \frac{1}{2}ma^2\omega^2\cos^2(\omega t + \theta) - \frac{1}{2}ma^2\omega^2 - mga\cos\omega t + mgl\cos\theta$

4.4 $H = \frac{p_r^2}{2m} + \frac{p_\theta^2}{2m} - \frac{GMm}{r}$

4.5 $H = \frac{p_\theta^2}{2ml^2} + \frac{p_\phi^2}{2ml^2\sin^2\theta} - mgl\cos\theta$

4.9 $H = \frac{p^2}{2m} + \frac{kz^2}{2} - kzz_0(t) - mgz$

5.1 (b) $F = -Q \ln(\cos q) + Q \ln Q - Q$

5.3 Not canonical if $K = H$.

5.4 $Q = p, P = p - 2q$

5.7 Evaluate $[H, [f, g]]$.

6.1 $S = \int \sqrt{\frac{1}{c^2} \left(\frac{k}{r} - E\right)^2 - m^2 c^2 - \frac{\alpha}{r}} + \sqrt{\alpha}\phi - Et$

6.2 $\frac{1}{2m} \left(\frac{\partial S}{\partial r}\right) + a + \frac{1}{2mr^2} \left[\left(\frac{\partial S}{\partial \theta}\right)^2 + 2mb\right] - E = 0$

Index

Printed in the United States
By Bookmasters